THE IMPACT OF BIOTECHNOLOGY ON AUTOIMMUNITY

Medical Science Symposia Series

Volume 6

The titles published in this series are listed at the end of this volume.

The Impact of Biotechnology on Autoimmunity

Edited by

A. G. Dalgleish

Department of Cellular and Molecular Sciences, St. George's Hospital, London, U.K.

A. Albertini

School of Medicine, University of Brescia, Brescia, Italy

and

R. Paoletti

Institute of Pharmacological Sciences, University of Milan, Milan, Italy

KLUWER ACADEMIC PUBLISHERS

DORDRECHT / BOSTON / LONDON

Fondazione Giovanni Lorenzini, Milan, Italy

Giovanni Lorenzini Medical Foundation, Houston, U.S.A.

A C.I.P. Catalogue record for this book is available from the Library of Congress

ISBN 0-7923-2724-1

Published by Kluwer Academic Publishers,
P.O. Box 17, 3300 AA Dordrecht, The Netherlands.

Kluwer Academic Publishers incorporates
the publishing programmes of
D. Reidel, Martinus Nijhoff, Dr W. Junk and MTP Press.

Sold and distributed in the U.S.A. and Canada
by Kluwer Academic Publishers,
101 Philip Drive, Norwell, MA 02061, U.S.A.

In all other countries, sold and distributed
by Kluwer Academic Publishers Group,
P.O. Box 322, 3300 AH Dordrecht, The Netherlands.

Printed on acid-free paper

Printed in the Netherlands

CONTENTS

PREFACE

The term "autoimmunity" has been used to categorize a number of different diseases of unknown etiology. The term as applied to many of these diseases would probably be interpreted best as "autoreactivity," as the clinical phenotypes are often characterized by an inflammatory-like accumulation of activated cells of the immune system at sites of obvious damage of normal cells and/or tissue. The reasons as to why an immune system should attack itself are far from clear, although the association with specific infectious diseases in genetically susceptible people remains perhaps our best lead. The input of the biotechnological revolution has enabled us to attempt to readdress many of the fundamental questions raised by clinical and serological associations with autoimmune disease. The ability to dissect the immune response to these infectious agents which are associated with autoimmune features (as well as the facility to identify new agents, e.g. HIV), in addition to the ability to clone and sequence immune response genes, has enabled a much better understanding, at least of the complexity of "autoimmunity" to be gleaned.

This volume contains the chapters that summarize the plenary presentations given at **The Impact of Biotechnology on AUTOIMMUNITY** meeting in Florence, Italy in June 1993. They cover all aspects from pathogenesis to treatment. The association with infectious diseases and autoimmunity is comprehensively covered by David Isenberg who reviews major issues, such as the association of autoantibodies appearing after infectious disease and antibacterial antibodies associated with autoimmune disease. Malak Kotb explores the interaction between streptococcal infections and autoimmune sequelae at the molecular level and suggests that streptococcal proteins acting as superantigens can activate specific subpopulations of the T-cell repertoire, and that this could lead to the autoimmune sequelae of streptococcal infection. HIV infection is known to be associated with many different serological and clinical features of autoimmune disease; several authors have proposed that it too may stimulate the immune system by acting as a superantigen. Angus Dalgleish and his colleagues suggest that the several regions of HLA homology within HIV may enable HIV to stimulate the immune system in a similar yet subtly different manner, in that chronic stimulation of the immune system by allogeneic MHC Class II results in a similar disease to that seen in HIV infection, in genetically susceptible mice. Although theoretical, a number of mechanisms are being pursued, including: the possibility that HIV envelope may functionally mimic HLA when expressed on an antigen presenting cell, the possibility that HIV HLA mimicking sequences (? Class II) may render self HLA (Class I) to appear as allogeneic to a syngeneic T-cell receptor, as well as the ability of HIV to incorporate HLA molecules into its envelope as it buds from cells. These last two possibilities could also apply to other immunopathogenic organisms (including many chronic infections such as leprosy, Leishmaniasis, etc.) and other enveloped viruses that bud from mammalian cells. In addition to the foregoing, the ability to mimic HLA and hence self epitopes, could allow a highly variable virus to escape an effective immune response, by acting as a decoy. The carboxy terminus of HIV which contains HLA-like sequences is immunodominant. No attempt at understanding autoimmunity would be complete without looking at the possible role of heat shock proteins (hsp). Here, Stefan Kaufmann reviews the role of the T-cell response to the various members of the hsp family. The ability of hsp to present peptides to the immune system and induce an antitumor response suggests that they are capable of functionally "aping" HLA and may even be a primitive forerunner (of HLA) so that primitive multicellular organisms could identify self from nonself cells. The implications for autoimmunity if true would be fundamental.

Autiommunity is associated with autoantibodies to various self proteins, especially nuclear proteins. However, the presence of antibodies does not always correlate with diseases even though specific antibodies tend to serve as diagnostic markers for specific diseases such as those of the connective tissues. Bombardieri and colleagues review the ability to detect these antibodies and their possible relevance. Specific antibodies such as Rheumatoid Factor (RF) binding the Fc region of IgG molecules are reviewed with regards to their relevance to disease and the other immune responses such as overproliferation of monocytes, macrophages, CD8 and CD4 cells. The association between antibody responses, vasculitis and diseases such as multiple myeloma and chronic acute liver disease is reviewed by Dammacco and colleagues. The strong genetic association with autoimmune diseases is the subject of several chapters which focus on diabetes in particular and coeliac disease. The association between the HLA Class II DQ gene and these diseases is striking, and an excellent example of how improvements in biotechnological techniques has led to a better understanding of the interaction of infectious agents with the hosts' immune system and the roles that different HLA types have to play in controlling that response. A detailed method to analyze this peptide level is reviewed by Sinigaglia et al. A full understanding of this could have important therapeutic implications.

Therapy has to be the ultimate goal of the biotechnological "war" on autoimmunity. A number of possibilities are reviewed such as the induction of peripheral tolerance in mice (Adorini et al.), the use of the anti-CD3 monoclonal antibodies (Chatenoud et al.) and the engineering of IL1 antagonists (Boraschi et al.). Chapters reviewing natural killer (NK) cell ontogeny and role of idiotypes in inducing autoimmunity also have potential therapeutic implications, as NK activity can be controlled with various cytokines and anti-idiotype immunizations can abrogate autoimmune-induced diseases in a number of rodent molecules.

This volume therefore covers the entire gamut of possibilities by which biotechnology may be used to help analyze and dissect autoimmune disease. In addition, biotechnology will provide the weapons by which to attack autoimmunity and hopefully turn off the chronic destructive processes that are all too evident in those millions of sufferers from such conditions as diabetes mellitus, rheumatoid arthritis, systemic lupus erythematosus, coeliac disease and psoriasis to name but a few. It is hoped that the approaches being taken to analyze "autoimmunity" reviewed here eventually will be developed into effective therapies for these conditions.

A.G. Dalgleish
A. Albertini
R. Paoletti

Acknowledgements

We would like to express our gratitude to the collaborators and colleagues at the Brescia Medical School and to the Lorenzini Foundation and its staff for their generous support and collaboration.

Angus George Dalgleish
Alberto Albertini
Rodolfo Paoletti

ROLE OF HLA-DQ2 AND -DQ8 AS SUSCEPTIBILITY MOLECULES IN COELIAC DISEASE - A MODEL-DISEASE OF AUTOIMMUNITY

Erik Thorsby, Knut E.A. Lundin and Ludvig M. Sollid
Institute of Transplantation Immunology
The National Hospital
Oslo
NORWAY

Coeliac disease (CD) is an immunological disorder of the small intestine, characterized by crypt-cell hyperplasia and villous atrophy, often leading to malabsorption and diarrhea. CD is precipitated in susceptible individuals by ingestion of gluten from wheat and similar proteins from rye and barley, and it is believed that an immune response to gluten (or rather gliadins in gluten) is essential in the immunopathogenesis. The disease also demonstrates strong HLA associations. Since CD shows many features in common with autoimmune diseases, such as the immunopathology and strong HLA associations, and since the triggering antigen is known (which is not the case for autoimmune diseases), it is a good model for studies of the role of HLA molecules in causing disease susceptibility. Here we summarize some recent studies demonstrating that two different HLA-DQ molecules, DQ(α1*0501, β1*0201), i.e. a variant of DQ2, and DQ(α1*0301, β1*0302), i.e. DQ8, confer the primary susceptibility to develop CD. Further we summarize some of our recent studies which indicate that the mechanism may involve preferential binding and presentation to $CD4^+$ T cells of given gluten-derived peptides by these DQ molecules in the intestinal mucosa.

CD is Primarily Associated to DQ2 (and DQ8)

Various HLA associations have been described in CD. In this disease, as in other HLA associated disorders, it is important first to establish the primary HLA association; i.e. to which of the many genes in the HLA complex is the disease primarily associated. Is the primary association to gene(s) encoding some of the classical HLA class I (A, B or C) or class II (DR, DQ or DP) cell-membrane molecules, or is the primary association instead to other genes in the HLA complex (LMP, TAP, etc.), in strong linkage disequilibrium to those encoding the classical HLA molecules?

A. G. Dalgleish et al. (eds.), The Impact of Biotechnology on Autoimmunity, 1–6.

A strong association of CD to DR3 has been found in all populations so far studied (see [1] for references). An increase of DR7 has also been found among CD patients, but DR7 seems to be associated with CD only in combination with DR3 or DR5. Since the same DQA1 gene (DQA1*0501) is found on the haplotypes encoding DR3 or DR5, and since the DQB1 gene carried by DR3 and most DR7 haplotypes is also identical (DQB1*0201), we suggested [2] that the primary association in CD was to a particular variant of the DQ2 molecule; i.e. that encoded by DQA1*0501 and DQB1*0201 in *cis* position on DR3 haplotypes, or in *trans* position in DR5/DR7 heterozygotes (see Figure 1). In other experiments we found that this DQ heterodimer (later called DQ2 for short) was expressed in the cell membrane both when encoded in *cis* or *trans* position [3]. A primary association of CD to this DQ heterodimer has later been corroborated in several studies (references in [1]), and in most populations investigated approximately 90% or more of CD patients carry this heterodimer (compared to 12-28 % among healthy controls). Associations to some DP alleles have also been described, but there now seems to be general agreement that this is due to linkage disequilibrium between these DP alleles and the CD associated DQ genes (see [1] for references). No associations to given TAP1 alleles have been found [4]. A primary association of CD to the DQ(α1*0501,β1*0201) heterodimer thus explains previously observed HLA associations in CD by linkage disequilibrium [1].

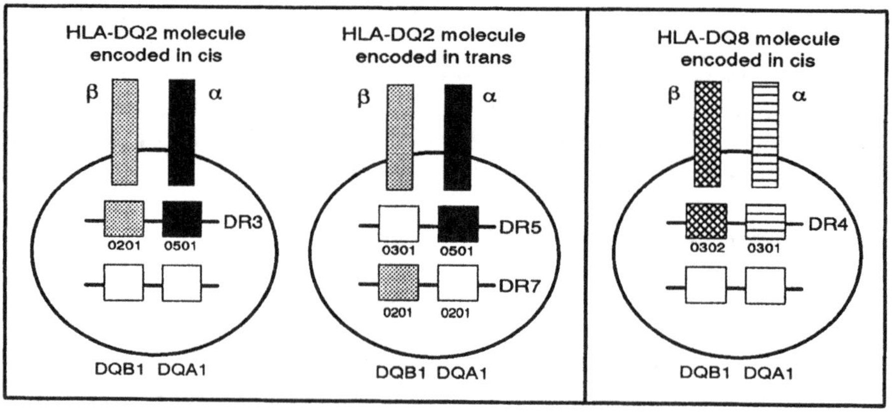

Figure 1. Left part: The DQ(α1*0501,β1*0201) variant of DQ2 is encoded by DQA1*0501 and DQB1*0201 in *cis* position on DR3 haplotypes and in *trans* position by DR5/DR7 heterozygotes.
Right part: The DQ8 molecule encoded by some DR4 haplotypes.

Approximately 2-10% of CD patients (approximately 20% of CD patients from Bologna) do not carry DR3, nor are they DR5/DR7 heterozygous (references in [1]). Some may carry the DQA1*0501 and DQB1*0201 genes on "unusual" haplotypes [5], most others are DR4 positive. The DR4 positive CD patients carry different variants of DR4, but almost without exceptions they all carry DQ8, (i.e. the DQ heterodimer encoded by DQA1*0301 and DQB1*0302; see Figure 1) and not DQ7 (which is also encoded by some DR4 haplotypes) [5,6].

It may therefore be concluded that the majority of CD patients carry DQ2 (encoded by DQA1*0501 and DQB1*0201 in *cis* or *trans* position), and that the relatively few CD patients who do not carry DQ2 instead carry DQ8. There are very few CD patients who do not carry DQ2 or DQ8. The susceptibility associated with DQ2 (or DQ8) seems to be dominant in that only one dosage of the corresponding pair of DQ genes is necessary to confer susceptibility. However, the penetrance apparently is low in that only a small proportion of those carrying DQ2 or DQ8 ever develop CD. This may be explained by an influence also of other (non-HLA) genes on CD susceptibility, environmental factors (in addition to ingestion of given cereal proteins), or both. The nature of these additional genetic or environmental factors is presently not known.

The CD Associated DQ2 and DQ8 Molecules in the Intestinal Mucosa may Preferentially Present Gluten-Derived Peptides to CD4$^+$ T Cells

The studies summarized above strongly suggest that susceptibility to develop CD is associated to the DQ2 or DQ8 molecules as such (and not to products of closely linked genes in the HLA complex). The next question is then where and how DQ2 and DQ8 may exert their effects. Possible mechanisms may involve thymic or extra-thymic effects of these DQ molecules. In the thymus DQ2 and DQ8 may positively select CD4$^+$ T cells with a particular repertoire of T cell receptor (TCR) genes, which under given circumstances may be activated by gluten-derived peptides, or DQ2 and DQ8 may negatively select (delete) T cells having a regulatory effect upon T cell activation by gluten-derived peptides. Alternatively, DQ2 and DQ8 of antigen-presenting cells (APC) in the intestinal mucosa may preferentially bind and present to CD4$^+$ T cells potential immunopathogenic peptides derived from gluten. These mechanisms are not mutually exclusive.

To study the latter possibility we challenged biopsies from the small intestine of CD patients on a gluten-free diet with a peptic/tryptic digest of gluten, whereby a proportion of CD4$^+$ T cells in the biopsies became activated and expressed CD25 (IL-2 receptor L chain) [7]. These T cells activated by gluten were then immunomagnetically isolated (using anti CD25 coated Dynabeads), and the activated T cells progagated using PHA and IL-2. T cell lines (TCL) and T cell clones (TCC) were established, and their specificity with respect to antigen and antigen-presenting HLA molecules was studied [8].

The results of some experiments are illustrated in Figure 2. Polyclonal gluten-specific TCL from three different CD patients responded strongly when gluten was presented by APC carrying the haplotype encoding their CD associated DQ heterodimers (DQ2 for patients CD280 and CD281, DQ8 for patient CD282), and this response was almost abolished using an anti-DQ Mab, while an anti-DR Mab had no effects. Thus, most CD4$^+$ T cells in the intestinal mucosa of these patients recognized gluten-derived peptides when presented by their CD associated DQ molecules, and not when presented by their DR molecules [8].

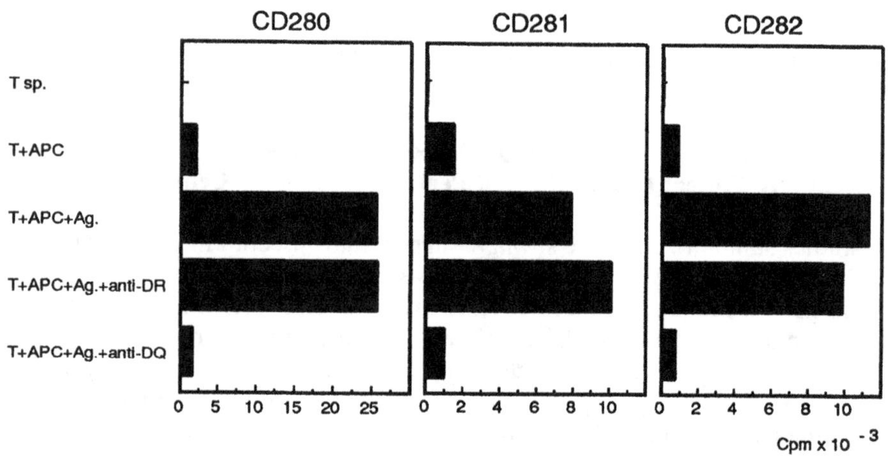

Figure 2. Polyclonal T cell lines from three CD patients (CD280:DR3/6,DQ1/2; CD281:DR3,DQ1/2; CD282:DR4,DQ7/8) were tested against gluten (=Ag) in the absence of any Mab, or in the presence of anti-DR or anti-DQ Mabs. As APC were used cells homozygous for DR3,DQ2(CD 280 and 281) or homozygous for DR4,DQ8(CD282).

The TCL and TCC from these and other patients were further studied using a panel of APC expressing different combinations of DR, DQ and DP molecules. The general finding was that a preponderance of the CD4$^+$ T cells from the intestinal mucosa of CD patients carrying DQ2 recognized gluten-derived peptides when presented by DQ2 (encoded by DQA1*0501 and DQB1*0201 either in *cis* or *trans* position). A gluten-specific response was rarely seen when gluten-derived peptides instead were presented with other DQ, DR or DP molecules carried by the patient. Similarily, when instead CD4$^+$ T cells from the intestinal mucosa of a CD patient being DR4;DQ7,DQ8 were investigated, a preponderance of them were found to recognize gluten-derived peptides presented by DQ8, and not by DQ7 nor DR4 (Lundin et al.; to be published). In contrast, when instead CD4$^+$ T cells from the peripheral blood of CD patients were activated with gluten, both DR, DQ and DP molecules were found to function as antigen-presenting molecules (Gjertsen et al., to be published).

Thus, a major proportion of gluten-specific T cells in the intestinal mucosa of CD patients recognizes gluten-derived peptides when presented by DQ2 or DQ8 (depending on which DQ molecules the CD patient is carrying). The TCR genes of some TCC were also sequenced, but no predominant TCR usage was found.

The most straight-forward explanation for these findings would be preferential presentation of given gluten-derived peptides to CD4$^+$ T cells by DQ2 or DQ8 in the intestinal mucosa. It is difficult to explain, however, why a similar preferential presentation of gluten-derived peptides by these DQ molecules was not observed using CD4$^+$ T cells from the peripheral blood. One possibility may be a higher expression of DQ molecules in APC in the intestinal mucosa compared to APC in the peripheral blood. Alternatively, the peptide repertoire following digestion of gluten in the intestinal lumen or by APC in the mucosa may differ from that generated by digestion in APC from peripheral blood. In any case, the T cell immune response to only some gluten-derived peptides may be immunopathogenic, peptides which are preferentially presented by DQ2 or DQ8.

To study this further we have recently established an assay for the detection of peptide-binding to DQ2 (Johansen et al.; to be published). We have found that different sets of peptides bind to DQ2 compared to DR3. Experiments are underway to establish the motif of peptides binding to DQ2, and to study which peptides derived from gluten might carry this motif and preferentially bind to DQ2 (or DQ8).

Conclusion

Available evidence strongly suggest that susceptibility to develop CD is primarily associated to the DQ heterodimers DQ2 and DQ8. Our studies also indicate that a preferential presentation of given immunopathogenic gluten-derived peptides to CD4$^+$ T cells by these DQ molecules in the intestinal mucosa is a possible mechanism.

However, several important questions remain to be answered. First, what are the mechanisms behind the tissue injury? We are currently investigating the cytokine profile of the DQ2 and DQ8 restricted gluten-specific T cells from the intestinal mucosa of CD patients. Secondly, why is it only peptides from gluten, and not those derived from many other ingested proteins, which in susceptible individuals may cause the immunopathology leading to CD? Is it possible that an autoimmune response may also contribute in CD; i.e. that given gluten-derived peptides via molecular mimicry also may activate T cells reacting with certain "self-antigens" in the intestinal mucosa? The fact that patients when treated on a gluten-free diet rather quickly recover would speak against an autoimmune component, but a prompt effect of given regulatory T cells (when large doses of gluten is avoided) cannot entirely be discarded. Thirdly, is it possible that only DQ2 and DQ8 may bind potential immunopathogenic peptides from gluten in the intestinal mucosa, and if so, why is it that only a minor proportion of individuals carrying DQ2 or DQ8 ever develop CD?

These and other questions are common or relevant for most autoimmune disorders. As mentioned, CD is a good model for further studies since the primary HLA associations have been established and the triggering antigen is known.

Acknowledgements

Our studies are supported by grants from the The Research Council of Norway and by Pronova AS.

References

1. Sollid LM, Thorsby E. HLA susceptibility genes in celiac disease. Genetic mapping and role in pathogenesis. Gastroenterology 1993; in press.
2. Sollid L, Markussen G, Ek J, Gjerde H, Vartdal F, Thorsby E. Evidence for a primary association of celiac disease to a particular HLA-DQ α/β heterodimer. J Exp Med 1989;169:345-50.
3. Lundin KEA, Sollid LM, Qvigstad E, et al. T lymphocyte recognition of a celiac disease-associated cis- or trans-encoded HLA-DQ α/β-heterodimer. J Immunol 1990;145:136-39.
4. Colonna M, Bresnahan M, Bahram S, Strominger JL, Spies T. Allelic variants of the human putative peptide transporter involved in antigen processing. Proc Natl Acad Sci USA 1992;89:3932-36.
5. Spurkland A, Sollid LM, Polanco I, Vartdal F, Thorsby E. HLA-DR and -DQ genotypes of celiac disease patients serologically types to be non DR3 or non DR5/7. Human Immunol 1992;35:188-192.
6. Mantovani V, Corazza R, Bragliani M, Frisoni M, Zaniboni MG, Gasbarrini G. Asp57-negative HLA DQβ chain and DQA1*0501 allele are essential for the onset of DQw2-positive and DQw2-negative coeliac disease. Clin Exp Immunol 1993;91:153-56.
7. Halstensen TS, Scott H, Farstad IN, Michaelsen TE, Brandtzaeg P. In situ two- and three-colour immunofluorescence staining of mucosal T cells in coeliac disease. Prog Histochem Cytochem 1992;26:201.
8. Lundin KEA, Scott H, Hansen T, et al. Gliadin-specific, HLA-DQ(α1*0501,β1*0201) restricted T cells isolated from the small intestinal mucosa of celiac disease patients. J Exp Med 1993; in press.

IMMUNOGLOBULIN V GENE REPERTOIRE OF HUMAN B CELLS

Robert S. Schwartz, MD
Department of Medicine
New England Medical Center
Tufts University School of Medicine
Boston, MA 02111

A notable feature of B cells that has been observed recently in several laboratories is that a relatively small group of V_H segments predominates in the expressed V_H gene repertoire of normal humans. The results of a number of different experiments suggest that the phenomenon of repertoire restriction is due to clonal selection at the level of both mature and immature B cells. For instance, in examining sequences of the over-represented V_H genes we found examples of somatic mutations that were typical of an antigen-driven immune response, and which were manifestations of selection of mature B cell clones [1]. But we also found cases of over-represented V_H genes that had germline sequences. Each of these unmutated V_H segments was linked to a clonally unique and unmutated (D-J_H) CDR3 [2]. These latter findings suggest a process of positive selection that is linked to the heavy chain in B cells that do not mutate their V_H genes. These and related results have led us and others to propose that clonal selection occurs not only in mature B cells but also in pre-B cells [3-5]. We propose that positive selection of pre-B cells differs from clonal selection of mature B cells in two ways: (a) pre-B cells lack a mechanism for V gene somatic mutation; and (b) positively selected pre-B cells have a differentiation advantage, whereas antigen-driven mature B cells gain a proliferative advantage. Analogous types of clonal selection are well-established in T cells. Mature T cells undergo clonal expansion when stimulated by an antigen, whereas immature T cells differentiate but do not proliferate when positively selected [6-11].

This hypothesis has another parallel with known properties of T cells in that it links B cell differentiation to the anti-self binding properties of the pre-B cell's antigen receptor. Positive (or negative) selection of immature T cells requires binding of the T cell receptor to an MHC-peptide complex displayed by an antigen-presenting cell in the thymic cortex (reviewed in ref. [12]). In the normal fetus, self proteins are the sole source of such peptides, except for any maternal proteins that might leak through the placenta. Indeed, chemical analysis has shown that the peptides that are constitutively bound to MHC molecules originate from intracellular self-proteins [13-19]. Whether the immature T cell undergoes positive

A. G. Dalgleish et al. (eds.), The Impact of Biotechnology on Autoimmunity, 7–10.

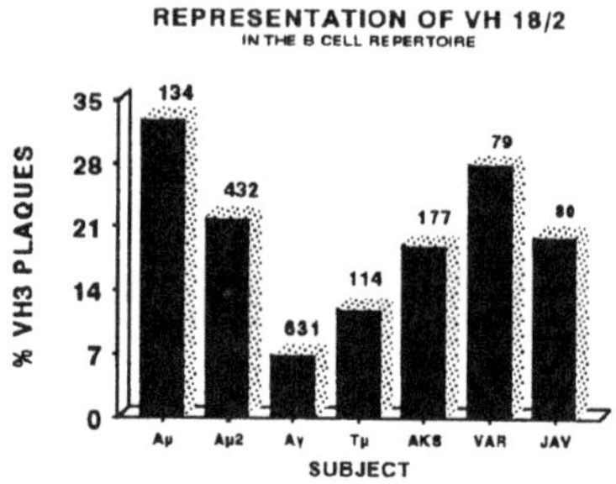

Figure 1. Representation of a single V_H gene ($V_H18/2$) in mature B cells from the peripheral blood of normal adults. (From Stewart, et al. J Exp Med 1992;177:1227.

or negative selection probably depends on the affinity of its receptor for the MHC + self-peptide complex displayed in the thymic cortex. It has been proposed that a high affinity receptor-MHC interaction kills the immature T cell, whereas low affinity binding promotes its differentiation [20]. The consequence of this arrangement is that endogenous peptides and self-reactive T cell receptors mold the naive T cell repertoire [21].

It is likely that all IgM antibodies encoded by germline variable region segments are low-affinity polyreactive autoantibodies (reviewed in refs. [22-24]). We propose that pre-B cells at the surface μ^+/light chain$^-$ stage complete their differentiation if their surface Ig heavy chains bind to self-antigens. It is likely that the surface μ chain transduces signals that instruct the pre-B cell to continue their differentiation. For example, expression of a function μ chain stimulates differentiation in IL-7 dependent immature mouse B cells; when the μ^+, IL-7 dependent B cells express a functional light chain they lose their dependence on IL-7 [25]. Indeed, targeted disruption of the membrane exon of the μ chain gene resulted in a severe deficiency of mature B cells, although there were normal numbers of pre-B cells in the marrow [26]. Furthermore, clonal expansion of marrow pre-B cells in the mouse probably depends on expression of the heavy chain [5].

In our hypothesis, a pre-B cell that expresses a polyreactive anti-self heavy chain would have an advantage, whereas a pre-B cell that expresses a μ heavy chain lacking those properties would have but a small possibility of moving to the next step in differentiation, and may, as a result, die. Certain V_H genes may therefore appear frequently in the expressed pre-immune B cell repertoire because the heavy chain region they encode allows binding to multiple autoantigens.

The light chain has an important role in specifying the antigen-binding properties of the immunoglobulin molecule. The light chain may thus also influence clonal selection, and hence the representation of a V_H gene in the expressed B cell repertoire. Our studies have not shown any linkage of $V_H18/2$, an over-represented V_H gene, to a particular light chain [27]. However, possible influences of the light chain on the frequency of a V_H gene in the expressed repertoire has not been systematically investigated.

References

1. Huang CC, Stewart AK, Schwartz RS, Stollar BD. Immunoglobulin heavy chain gene expression in peripheral blood B-lymphocytes. J Clin Invest 1992;89:1331-1343.
2. Stewart K, Huang CC, Stollar BD, Schwartz RS. High frequency representation of a single VH gene in the rearranged and expressed human B cell repertoire. J Exp Med 1992;in press.
3. Schwartz, RS. The human B cell repertoire: Is repertoire bias a manifestation of life saver heavy chain genes? Festschrifft For Professor Ivan Roitt 1993;in press.
4. Nishimoto N, Kubagawa H, Ohno T, Gartland GL, Stankovic AK, Cooper MD. Normal pre-B cells express a receptor complex of μ heavy chains and surrogate light-chain proteins. Proc Natl Acad Sci (USA) 1991;88:6284-6288.
5. Decker DJ, Boyle NE, Koziol JA, Klinman NR. The expression of the Ig H-chain repertoire in developing bone marrow B-lineage cells. J Immunol 1991;146:350-361.
6. Kisielow P, Bluthmann H, Staerz UE, Steinmetz M. Tolerance in T-cell-receptor transgenic mice involves deletion of nonmature CD4+CD8+ thymocytes. Nature 1988;333:742-746.
7. Macdonald HR. Mechanisms of immunological tolerance. Science 1989;246 (4933):982.
8. Mueller DL, Jenkins MK, and Schwartz RH. Clonal expansion versus functional clonal inactivation. Ann Rev Immunol 1989;7:45-80.
9. Schwartz RH. Acquisition of immunologic self-tolerance. Cell 1989;57:1073-1081.
10. Tomonari K. Tolerance in vivo and in vitro. Immun Rev 1990;116:139-157.
11. Von Boehmer H, Kisielow P. Self-nonself discrimination by T-cells. Science 1990;248:1369-1373.
12. Von Boehmer H. Developmental biology of T cells in cell receptor transgenic mice. Ann Rev Immunol 1990;8:531-556.
13. Buus S, Sette A, Colon SM. Autologous peptides constitutively occupy the antigen binding site of Ia. Science 1988;242:1045-1047.
14. Chicz RM, Urban RG, Lane WS, et al. Predominant naturally processed peptides bound to HLA-DR1 are derived from MHC-related molecules and

are heterogeneous in size. Nature 1992;358:764-768.
15. Hunt DF, Henderson RA, Shabanowitz J, et al. Characterization of peptides bound to the class-I MHC molecule HLA-A2.1 by mass spectrometry. Science 1992;255:1261-1263.
16. Madden DR, Gorga JC, Strominger JL, Wiley DC. The structure of HLA-B27 reveals nonamer self-peptides bound in an extended conformation. Nature 1991;353:321-325.
17. Murphy DB, Rath S, Pizzo E, et al. Monoclonal antibody detection a major self peptide: MHC class-II complex. J Immunol 1992;148:3483-3491.
18. Rudensky AY, Prestonhurlburt P, Hong SC, Barlow A, Janeway CA. Sequence analysis of peptides bound to MHC class-II molecules. Nature 1991;353:622-627.
19. Srinivasan M, Marsh EW, Pierce SK. Characterization of naturally processed antigen bound to major histocompatibility complex class-II molecules. Proc Natl Acad Sci (USA) 1991;88:7928-7932.
20. Sprent J, Gao EK, Webb SR. T-cell reactivity to MHC molecules - Immunity versus tolerance. Science 1990;248:1357-1363.
21. Pullen AM, Marrack P, Kasppler JW. The T-cell repertoire is heavily influenced by tolerance to polymorphic self-antigens. Nature 1988;335:796-801.
22. Stewart AK, Huang C, Long AA, Stollar BD, Schwartz RS. VH gene representation in autoantibodies reflects the normal human B cell repertoire. Immunol Rev 1992;in press.
23. Schwartz RS. Autoantibodies and normal antibodies: Two sides of the same coin. Harvey Lect 1987;81:53.
24. Schwartz RS. Polyvalent anti-DNA autoantibodies: Immunochemical and biological significance. Int Rev Immunol 1988;3:97-115.
25. Era T, Ogawa M, Nishikawa SI, et al. Differentiation of growth signal requirement of B lymphocyte precursor is directed by expression of immunoglobulin. EMBO J 1991;10:337-342.
26. Kitamura D, Roes J, Kuhn R, Rajewsky K. A B cell-deficient mouse by targeted disruption of the membrane exon of the immunoglobulin μ chain. Nature 1991;350:423-426.
27. Young F, Tucker L, Rubinstein D, et al. Molecular analysis of a germ line encoded idiotypic marker of pathogenic human lupus autoantibodies. J Immunol 1990;145:2545-2553.

RECENT PROGRESS IN HUMAN NATURAL KILLER CELL ONTOGENY

Maria Cristina Mingari*^, Alessandro Poggi*, Simona Sivori, Nicoletta Pella*, Chiara Vitale*, Roberto Biassoni*, and Lorenzo Moretta*°
**Istituto Nazionale per la Ricerca sul Cancro*
Genova
^Istituto di Oncologia Clinica e Sperimentale
Università di Genova
°Dipartimento di Medicina ed Oncologia Sperimentale
Università di Torino
sede di Novara
ITALY

Introduction

Natural killer (NK) cells substantially differ from T or B lymphocytes since they do not express (or rearrange) known receptors for antigen, namely surface Ig and the CD3/TCR complex [1,2]. They are characterized by the expression of low affinity surface receptors for IgG (FcγR III or CD16) [3] and by the N-CAM-homologous CD56 molecules [4-8]. NK cells display a size larger than resting T or B lymphocytes and typical cytoplasmic azurophilic granules. Because of these morphological features they have been termed "large granular lymphocytes" (LGL) [9]. NK cell maturation can occur in the absence of a functional thymus [10]. Indeed, functional NK cells are present in nude mice as well as in severe combined immunodeficiency (SCID) mice [11,12]. Their name originates from the fact that, different from resting T or B lymphocytes, NK cells are spontaneously able to mediate their effector function (i.e. with no need for prior activation *in vitro* or *in vivo*). In general, NK cells have been considered as nonclonally distributed effector cells that lyse tumor or virally infected cells via "nonspecific" mechanisms [1].

Several of the generally accepted concepts on NK cells have been recently challenged by evidence based on a variety of new experimental approaches and technological breakthroughs. In this context, in humans, important progress has been possible since cloning of NK [13,14] cells has become available and applied to the study of both functional capabilities and surface molecules with receptor function. Thus, based on the clonal approach, a number of surprising results have been obtained which have substantially modified our view on NK cells. Most of these results [15-21] have been recently reviewed [22,23] and will not be discussed

A. G. Dalgleish et al. (eds.), The Impact of Biotechnology on Autoimmunity, 11–19.

here. We shall here summarize and discuss some of our new data on the NK cell origin.

Origin of NK Cells

Commitment to the B or T cell lineage is characterized by an irreversible rearrangement of the immunoglobulin or TCR genes, respectively. In contrast, NK cell do not rearrange Ig or TCR genes [24-26] and can undergo phenotypic and functional maturation also in SCID mice [12]. In addition, they undergo normal development in the absence of a functional thymus [11]. In spite of the "thymus independence" and of the lack of a TCR rearrangement, NK cells share with T cells a number of properties regarding the expression of surface markers and effector functions including cytolytic activity and lymphokine production [1]. These data have supported the notion that T and NK cells may belong to related lineages.

Recent data may help to understand better the relationship existing between the two cell lineages. These data have been obtained by the analysis of lymphocyte differentiation during fetal ontogeny, as well as by the development of *in vitro* systems which mimic *in vivo* the environments necessary for NK or T cell maturation from precursors present either in human postnatal thymus or in fetal liver.

In mice a population of early fetal thymocytes has the potential to differentiate towards T or NK cell lineages depending on the *in vivo* microenvironment [27]. Thus, a predominant thymic population has been detected at 14.5 days of gestation which lacks both CD4 and CD8 differentiation antigens, but express FcγR II/III several days before the expression of TCR. If this population was maintained in a thymic microenvironment it underwent differentiation towards mature T cells. On the other hand, when removed from the thymus, this population selectively generated functional NK cells *in vivo* as well as *in vitro*. These data clearly indicate that a typical marker of mature NK cells is expressed early during T cell ontogeny. More importantly, they indicate that an immature thymocyte population gives rise to T or NK cells depending on the microenvironment [27].

Previous research *in vitro* performed in humans showed that immature thymocytes (isolated during cardiac surgery from patients of age ranging from 2 months to 4 years) characterized by the CD7$^+$3$^-$4$^-$8$^-$ phenotype undergo *in vitro* maturation towards T or NK cell lineages [28-33]. Again, the pattern of maturation was strictly dependent on the type of microenvironment and stimuli provided. Thus, growth of CD3$^-$16$^+$ cells with strong non-MHC restricted cytolytic activity was obtained both at the population and at the clonal level in the presence of exogenous IL-2 and H9 leukemia cell line as a source of irradiated feeder cells [30]. Since over 50% of clones obtained under limiting dilution conditions were phenotypically and functionally undistinguishable from typical NK cells clones derived from peripheral blood, these data clearly indicated that precursors of NK cells are present within human thymus. A possible explanation for the H9-cell dependence of the phenomenon may be the involvement of cytokines and/or cell surface molecules expressed by H9 cells and active on NK cell maturation/proliferation. Thus, H9 cells would mimic the microenvironment required for the physiological NK cell development. Along this line, proliferation of $CD16^+$ cells from immature thymocytes has been obtained in the presence of thymic stromal cells as a source of feeder cells [31]. When $CD3^-4^-8^-$ thymocyte populations were cultured in the presence of PHA and IL-2, gave rise to $CD3^+TCR^+$ mature T cell populations which expressed either γ/δ [32] or α/β [33] functional receptors.

A similar experimental approach was applied to the study of precursors present in the human liver isolated from embryos or fetuses of 6-10 weeks of gestation [34]. In the presence of H9 and IL-2, liver cells suspensions underwent both proliferation and maturation towards CD3-16+ NK cells (see schematic figure). Importantly, the precursors of NK cells were restricted to a small population expressing the common leukocyte antigen CD45 (2-15% of total cells recovered from Ficoll Hypaque gradients). These data indicated that precursors capable of *in vitro* differentiation towards NK cells can be identified as early as at 6 weeks of gestation. Unexpectedly, by applying to liver cell suspensions the "*in vitro*" "microenvironment" able to induce T cell maturation from thymic precursors, we could obtain cell proliferation and maturation towards $CD3^+/TCR\alpha/\beta^+$ or $CD3^+/TCR\gamma/\delta^+$ cell populations or clones (see figure). It should be noted that these data could be obtained also in embryos at 6-8 weeks of gestation, that is before colonization of the thymic rudiment by T cell precursors. The gestational age together with the finding that $CD3^+TCR^+$ cells could not be generated from cells isolated from the upper part of thorax of the same embryos, clearly indicate that liver-derived proliferating lymphocytes are not thymic emigrants, but they are rather derived from liver precursors. Again, the T cell precursors were confined to the small $CD45^+$ subset of liver cells [34]. It should be emphasized that these data provide the first experimental evidence that mature T cells can derive *in vitro* from early fetal liver precursors without a functional thymus and may provide insights in the mechanism of T cell maturation and selection. Interestingly, similar data have been reported by Sànchez et al., by using different culture systems [35].

Taken together, data in mouse and human thymocytes, as well as in human fetal liver indicate that these cell populations may contain precursors that are committed either to the T or to the NK cell lineage: if this is the case, appropriate maturation signals would be sufficient for these committed precursors to undergo maturation to T or NK cells. An additional explanation, which is equally compatible with available data, is that thymus and/or fetal liver contain uncommitted lymphoid precursor or pluripotent stem cells. This would imply that the culture conditions used *in vitro* would mimic the microenvironment (e.g. thymus) needed to induce differentiation of uncommitted precursors towards T or NK cells. Both data in mouse and in humans do not clarify whether T or NK cells are derived from the same or different cell precursors. It is clear that the solution to this problem would be possible only after the development of culture conditions that allow, in single cell assays, optimal maturation towards different hemopoietic cell lineages, including T and NK cells.

Expression of CD3 Genes during NK Cell Maturation

Although a major discrimination between mature T and NK cells is the rearrangement and the expression of a functional TCR/CD3 complex, nevertheless certain molecules of the CD3 complex are also expressed in NK cells [36]. Thus, CD3 ζ chains have been shown to be associated to CD16 [37], CD2 [38] or p58 [39] molecules (see below). In addition, transcripts for CD3 ε have been described in both polyclonal and clonal NK cell populations [40]. Recently, by the use of antisera raised against a CD3 ε deduced peptide sequence, Lanier and co-workers would detect the CD3 ε protein at the cytoplasmic level [41]. In no instances could transcripts for CD3 γ and δ be detected in resting or activated mature NK cells.

Analysis of the expression of the various CD3 chains at the level of embryonic/fetal NK cells showed that, in addition to ζ chain, also CD3 ε and CD3 δ chains and low amount of CD3 γ chain [41] could be detected at the cytoplasmic level. Moreover, analysis of large numbers of $CD3^-16^+56^+$ NK clones derived from immature thymic populations ($CD3^-4^-8^-$), under appropriate culture conditions (see above), revealed, in a portion of NK clones, cytoplasmic CD3 ε proteins as detected with conventional anti-CD3 mAb (specific for the CD3 ε chain) [21]. Thus, it appears that the precursors of NK cells express a set of CD3 proteins which is more closely related to mature T cells than to mature NK cells [21,41]. These findings suggest, but do not prove, that the precursors for the two cell lineages either share a common set of activated genes or are identical. In the latter case (i.e. if T and NK cells arise from a common precursor) while the TCR rearrangement in T cells would be accompanied by the expression of the full set of CD3 genes, maturation towards NK cells would be characterized by the down regulation of CD3 γ and δ genes [36].

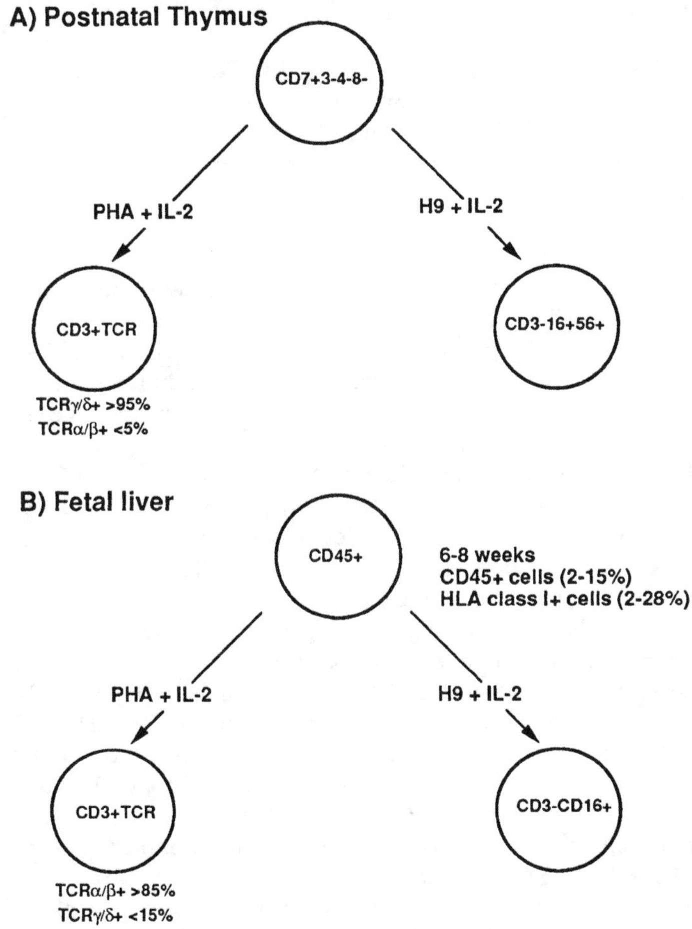

A) Human Immature Thymocytes ($CD7^+3^-4^-8$) contain precursors of both T and NK cells and their development towards T or NK cells depends on the *in vitro* or *in vivo* microenvironment.
Note that highly purified $CD7^+3^-4^-8^-$ cells, in the presence of PHA and IL-2, give rise mostly to TCR γ/δ^+ cells (only few, if any, $TCR\alpha/\beta^+$ cells can be detected). In the presence of H9 cells and IL-2, most cells are $CD3^-16^+56^+$ and display a strong cytolytic activity against both NK sensitive and NK resistant target cells.

B) Human Fetal Liver Cells contain precursors capable of maturation towards T or NK cell lineage *in vitro*. By applying the same culture conditions used for thymocytes, cell suspensions (purified on Ficoll-Hypague gradients) undergo maturation towards $TCR\alpha/\beta$ (>85%) or γ/δ (<15%). Note that the cell suspensions derive from fetuses prior to the colonization of epithelial thymic rudiment with lymphoid precursors, occurring between 8 and 9 weeks. This provides the first experimental evidence of an extrathymic *in vitro* T cell maturation.
In the presence of H9 and IL-2 growth of $CD3^-16^+56^+$ NK cells occurs.
Note that the precursors of both T and NK cells are confined to the small $CD45^+$ cell subset.

Acknowledgements

We wish to thank Mrs. Cinzia Miriello and Mr. Stefano Canu for their skillful assistance in the preparation of the manuscript. This work was supported, in part, by grants by the Consiglio Nazionale delle Richerche (Progetto A.C.R.O.), Associazione Italiana per la Ricerca sul Cancro, and Istituto Superiore di Sanità.

References

1. Trinchieri G. Biology of natural killer cells. Adv Immunol 1989;47:187.
2. Lanier LL, Phillips J, Hackett J, Tutt M, Kumar V. Natural killer cells: definition of a cell type rather than a function. J Immunol 1986;137:2375.
3. Kraft D, Lanier LL. The Fc receptor for IgG on human natural killer cells: phenotypic, functional and comparative studies using monoclonal antibodies. J Immunol 1984;133:180.
4. Griffin JD, Hercend T, Beveridge R, Schlossman SF. Characterization of an antigen expressed on human natural killer cells. J Immunol 1983;130:2947.
5. McGarry RC, Pinto A, Hammersley-Straw DR, Trevenen CL. Expression of markers shared between human natural killer cells and neuroblastoma lines. Cancer Immunol Immunother 1988;27:47.
6. Lanier LL, Chang C, Azuma M, Ruitenberg JJ, Hemperly JJ, Phillips JH. Molecular and functional analysis of human NK-cells associated neural cell adhesion molecule (N-CAM/CD56). J Immunol 1991;146:4421.
7. Lanier LL, Le AM, Civin CI, Loke MR, Phillips JH. The relationship of CD16 (Leu11) and Leu19 (NKH1) antigen expression on human peripheral blood NK cells and cytotoxic T lymphocytes. J Immunol 1986;136:4480.
8. Shubert W, Zimmermann K, Cramer M, Powitz S. Lymphocyte antigen Leu-19 as a molecular marker of regeneration in human skeletal muscle. Proc Natl Acad Sci USA 1989;86:307.
9. Saksela E, Timonen R, Ranki A, Häyry P. Morphological and functional characterization of isolated effector cells responsible for human natural killer activity to fetal fibroblasts and to cultured cell line targets. Immunol Rev 1979;44:71.
10. Herberman RB, Nunn MF, Lavrin DH. Natural cytotoxic reactivity of mouse lymphoid cells against syngenic and allogeneic tumors. II. Characterization of effector cells. Int J Cancer 1975;16:230.
11. Hackett JJ, Bosma GC, Bosma MJ, Bennet M, Kumar V. Transplantable progenitors of natural killer cells are distinct from those of T and B lymphocytes. Proc Natl Acad Sci USA 1986;83:3427.
12. Dorshkind K, Pollack SB, Bosma MJ, Phillips RA. Natural killer (NK) cells are present in mice with severe combined immunodeficiency (scid). J Immunol 1985;134:3798.
13. Moretta A, Pantaleo G, Moretta L, Mingari MC, Cerottini JC. Direct

demonstration of the clonogenic potential of every human peripheral blood T cells. Clonal analysis of HLA-DR expression and cytolytic activity. J Exp Med 1983;157:743.

14. Pantaleo G, Zocchi MR, Ferrini S, Poggi A, Tambussi G, Bottino C, Moretta L, Moretta A. Human cytolytic cell clones lacking surface expression of T cell receptor α/β or γ/δ. Evidence that surface structures other than CD3 or CD2 molecules are required for signal transduction. J Exp Med 1988;168:13.
15. Moretta A, Tambussi G, Bottino C, Tripodi G, Merli A, Ciccone E, Pantaleo G, Moretta L. A novel surface antigen expressed by a subset of human $CD3^-$ $CD16^+$ natural killer cells. Role in cell activation and regulation of cytolytic function. J Exp Med 1990;171:695.
16. Moretta A, Bottino C, Pende D, Tripodi G, Tambussi G, Viale O, Orengo AM, Barbaresi M, Merli A, Ciccone E, Moretta L. Identification of four subsets of human CD3-CD16+ NK cells by the expression of clonally distributed functional surface molecules. Correlation between subset assignment of NK clones and ability to mediate specific alloantigen recognition. J Exp Med 1990;172:1589.
17. Moretta A, Vitale M, Bottino C, Orengo AM, Morelli L, Augugliaro R, Barbaresi M, Ciccone E, Moretta L. P58 molecules as putative receptors for MHC class I molecules in human natural killer (NK) cells. Anti-p58 antibodies reconstitute lysis of MHC class I-protected cells in NK clones displaying different specificities. J Exp Med 1993, in press.
18. Ciccone E, Viale O, Pende D, Malnati M, Biassoni R, Melioli G, Moretta A, Long EO, Moretta L. Specific lysis of allogeneic cells after activation of CD3-lymphocytes in mixed lymphocyte culture. J Exp Med 1988 168:2403.
19. Ciccone E, Pende D, Viale O, Tambussi G, Ferrini S, Biassoni R, Longo A, Guardiola J, Moretta A, Moretta L. Specific recognition of human CD3-CD16+ NK cells requires the expression of an autosomic recessive gene on target cells. J Exp Med 1990;172:47.
20. Ciccone E, Pende D, Viale O, Di Donato C, Tripodi G, Orengo AM, Guardiola J, Moretta A, Moretta L. Evidence of Natural Killer (NK) cell repertoire for (allo)antigen recognition: definition of five distinct NK-determined allospecificities in humans. J Exp Med 1992;175:709.
21. Ciccone E, Pende D, Viale O, DiDonato C, Orengo AM, Biassoni R, Verdiani S, Amoroso A, Moretta A, Moretta L. Involvement of HLA class I alleles in NK cell specific function: expression of HLA-Cw3 convers selective protection from lysis by alloreactive NK clones displaying a defined specificity (specificity 2). J Exp Med 1992;176:963.
22. Moretta L, Ciccone E, Moretta A, Höglund P, Öhlen C, Karre K. Allorecognition by NK cells: nonself or no self? Immunol Today 1992;13:300.
23. Moretta L, Ciccone E, Mingari MC, Biassoni R, Moretta A. Human NK

cells: Origin, clonality, specificity and receptors. Advances in Immunology 1993, in press.

24. Herberman RB, ed. NK cells and other natural effector cells. Academic Press, New York, 1982.

25. Lanier LL, Cwirla S, Phillips JH. Genomic organization of T cell γ genes in human peripheral blood natural killer cells. J Immunol 1986;137:3375.

26. Lanier LL, Cwirla S, Federspiel N, Phillips JH. Human natural killer cells isolated from peripheral blood do not rearrange T cell antigen receptor β chain genes. J Exp Med 1986;163:209.

27. Rodewald HR, Moingeon P, Lucich JL, Dosiou C, Lopez P, and Reinherz EL. A population of early fetal thymocytes expressing FcγRII/III contains precursors of T lymphocytes and natural killer cells. Cell 1992;69:139.

28. Tutt M, Schuler W, Kuziel WA, Tucker PW, Bennet M, Bosma MJ, Kumar V. T cell receptor genes do not rearrange or express functional transcripts in natural killer cells of scid mice. J Immunol 1987;138:2338.

29. Poggi A, Biassoni R, Pella N, Paolieri F, Bellomo R, Bertolini L, Moretta L, and Mingari MC. In vitro expansion of CD3/TCR- human thymocyte populations that selectively lack CD3 gene expression: a phenotypic and a functional analysis. J Exp Med 1990;172:1409.

30. Mingari MC, Poggi A, Biassoni R, Bellomo R, Ciccone E, Pella N, Morelli L, Verdiani S, Moretta A, Moretta L. In vitro proliferation and cloning of $CD3^{-}$ $CD16^{+}$ cells from human thymocyte precursors. J Exp Med 1991;174:21.

31. De la Hera A, Marston W, Aranda C, Toribio ML, Martìnez-A C. Thymic stroma is required for the development of human T cell lineages in vitro. Internal Immunol 1989;1:471.

32. Mingari MC, Varese P, Bottino C, Melioli FG, Moretta A, Moretta L. Clonal analysis of CD4-CD8- human thymocytes expressing a T cell receptor γ/δ chain. Direct evidence for the de novo expression of CD8 surface antigen and of cytolytic activity against tumor targets. Eur J Immunol 1988;18:1831.

33. Blue ML, Levin H, Daley JF, Craig KA, Schlossman SF. Development of natural killer cells in human thymocyte culture: regulation by accessory cells, Eur J Immunol 1987;17:669.

34. Poggi A, Sargiacomo M, Biassoni R, Pella N, Sivori S, Revello V, Costa P, Valtieri M, Russo G, Mingari MC, Peschle C, Moretta L. Extrathymic differentiation of T lymphocytes and natural killer cells from human embryonic liver precursors. Proc Natl Acad Sci USA 1993;90:4465.

35. Sànchez MJ, Gutiérrez-Ramos JC, Fernàndez E, Leonardo E, Lozano J, Martìnez-A C, Toribio ML. Putative prethymic T cell precursors within the early human embryonic liver: a molecular and functional analysis. J Exp Med 1993;177:19.

36. Lanier LL, Spits H, Phillips JH. The developmental relationship between

NK cells and T cells. Immunol Today 1992;13:392.

37. Anderson P, Caligiuri M, O'Brien C, Manley T, Ritz J, Schlossman SF FcγRIII (CD16) is included in the ζ NK receptor complex expressed by human natural killer cells. Proc Natl Acad Sci USA 1990;87:2274.

38. Moingeon P, Lucich JL, McConkey DJ, Letourneur F, Malisse B, Kochan J, Chang HC, Rodewald HR, Reinherz EL. CD3 zeta dependence of the CD2 pathway of activation in T lymphocytes and natural killer cells. Proc Natl Acad Sci USA 1992;89:1492.

39. Moretta A, Bottino C, Pende D, Tripodi G, Orengo AM, Ciccone E, Moretta L. Subpopulations of human $CD3^-$, $CD16^+$ NK cells identified by MAb directed to triggering surface molecules: correlation with specific alloantigen recognition. In: NK cell mediated cytotoxicity: receptors, signaling and mechanism. R. Herberman, E. Lotzova (eds.) CRC Press, Boca Raton 1992, p. 431.

40. Biassoni R, Ferrini S, Prigione I, Moretta A, Long EO. CD3-negative lymphokine activated cytotoxic cells express the CD3ε gene. J Immunol 1988;140:1685.

41. Phillips JH, Hori T, Nagler A, Bhat N, Spits H, Lanier LL. Ontogeny of human natural killer (NK) cells: fetal NK cells mediate cytolytic function and express cytoplasmic CD3ε, δ proteins. J Exp Med 1992;75:1055.

ROLE OF THE HLA-DQ GENOTYPE IN IDDM SUSCEPTIBILITY

Roberto S. Accolla and Giovanni Tosi
Institute of Immunology and Infectious Diseases
School of Medicine, University of Verona
Policlinico Borgo Roma
37100 Verona
ITALY

Introduction

Insulin-dependent diabetes mellitus (IDDM) is the result of the selective destruction of pancreatic beta cells. The observation of insulitis with T and B lymphocytes infiltrating the islets of Langerhans during the acute phase of IDDM and the presence of islet-cell autoantibodies (ICA) indicate that IDDM is mediated by a specific autoimmune attack directed against cellular constituents of beta cells [1]. IDDM is a complex disease whose etiology involves both genetic or environmental factors [2]. Susceptibility to IDDM is certainly polygenic [3-4] and strongly correlates to genes mapping within the major histocompatibility complex, H-2 in the mouse and HLA in man. In man the strongest association was found in particular haplotypes of HLA class II region [5], a heterogeneous system encompassing at least three subregions designated DP, DQ and DR. Each subregion includes at least two expressible genes, A and B, coding for the two subunits, alpha and beta respectively, making up the class II cell surface glycoproteins [6]. These molecules are fundamental for the homeostasis of the immune system. They bind peptide antigens recognized by T helper cells, which play a key role in triggering an efficient cellular and humoral immune response. The autoimmune nature of IDDM and the genetic susceptibility linked to polymorphic molecules regulating the immune response have strongly stimulated molecular studies on the class II genes of patients to possibly correlate structural characteristics of given alleles to the disease susceptibility. Sequence studies and typing of HLA class II alleles by PCR and allele specific oligonucleotide (ASP) probes have shown that the alleles of IDDM patients are all present in normal controls. Moreover susceptibility and protection to IDDM, previously ascertained to be linked more to DQ than DP or DR alleles [7], were found to correlate in caucasoid population with codon 57 or DQB1 beta-2 exon which encodes the polymorphic beta-1 domain of the corresponding DQ subunit [8]. Protection by the presence of an Aspartic Acid residue (D) was

A. G. Dalgleish et al. (eds.), The Impact of Biotechnology on Autoimmunity, 21–26.

reported both in homozygous and heterozygous conditions whereas susceptibility correlated with the presence of a different residue in the same position (non-Aspartic = nD) that is a nD/nD homozygous genotype [8,9]. However, subsequent studies have demonstrated that, at least in the caucasoid population, presence of a D/nD heterozygous genotype was not associated to resistance [10,11]. Furthermore, recent studies have also indicated that DQA1 locus may contribute to the susceptibility trait. Codon 52 of the DQA1 alpha-2 exon encoding the polymorphic alpha-1 domain has been used as a marker for these studies. Alleles carrying an Arginine residue (R) in this position correlated with susceptibility [12]. In this study we summarize our recent studies on the HLA-DQ genetic analysis in 39 diabetic patients from Northeast Italy and 51 geographically matched unrelated normal controls. Molecular typing of DQB1 and DQA1 loci was performed by hybridizing PCR amplified DNA with allele-specific oligonucleotide probes (ASO). To the classical typing protocol using radioactive ASO probes and Southern blot analysis, we have added a recently introduced method, DEIA (DNA-enzyme immunoassay) [11]. In this technique a mouse monoclonal antibody specific for double stranded DNA is used to reveal the reaction between ASO probes, immobilized on plastic support, and PCR-amplified DNA. The anti ds-DNA mAb is revealed by adding anti mouse Igs labelled with chromogen. DEIA results are perfectly superimposable to those obtained by autoradiographic analysis.

Genotype Analysis of the DQ Locus.

We analyzed the DQ genotype in relation to the D or nD situation at residue 57 of DQB1 locus and to the R or nR situation at residue at 52 of the DQA1 locus. Results of such analysis are presented in Figure 1. Individuals with two DQB1 alleles having a nD residue in position 57 and two DQA1 alleles with a R residue in position 52 had the highest relative risk for disease (RR:72.3); they constituted 41% of IDDM patients as compared to 0% of controls. Heterozygosity either at residue 57 of DQB1 or residue 52 of DQA1 was sufficient to abrogate statistical significance for disease association. The majority of patients not included in the high risk DQB1 (nD)-DQA1(R) homozygous genotype were those in which one haplotype was completely susceptible and the other included either a susceptible DQB1 (28.2%) or a susceptible DQA1 (15.4%) allele. The precise allelic composition of the DQ genotype in IDDM patients analyzed in this study, reveals that statistical significance for susceptibility is reached only by two nD-R/nD-R genotypes: DQB1(0201)-DQA1(A4)/DQB1(0302)-DQA1(A3) and the complete homozygous DQB1(0201)-DQA1(A4)/DQB1(0201)-DQA1(A4). Based on the number of possible susceptible heterodimers an individual can form, it was found that 84.6% of IDDM cases could form 2 or more heterodimers (2 in cis and 2 in trans), but no IDDM cases was found to form 1 susceptible heterodimer in cis.

7.7% if IDDM patients and 47% of normal controls cannot express a susceptible heterodimer, either "in cis" or "in trans" configuration, and this phenotype gives a highly significant association to protection from disease (RR:0.1).

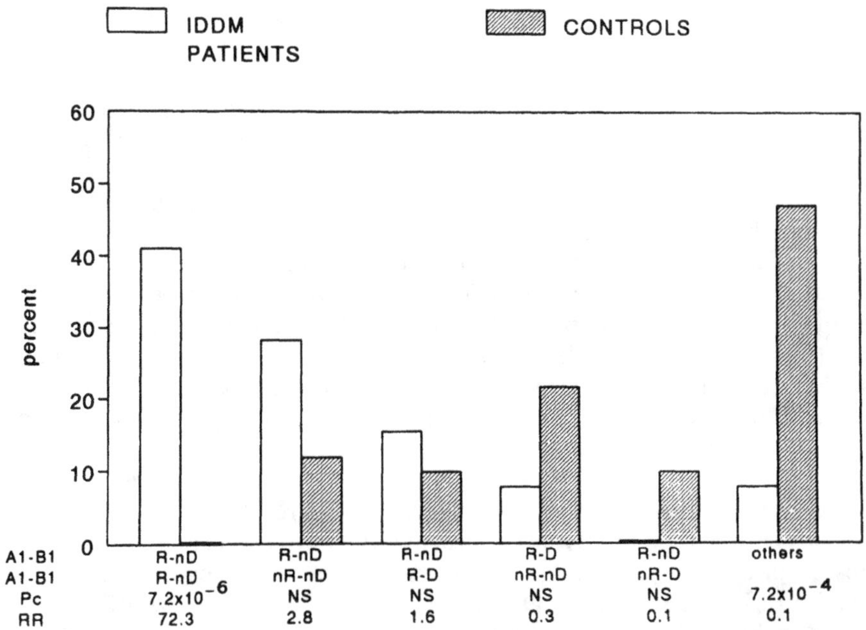

Figure 1. Distribution of DQ A1-B1 genotypes in IDDM patients (white columns) and controls (hatched columns). The possible genotypes, referred to the presence (D) or absence (nD) of Aspartic Acid residue in position 57 of DQB1 and to the presence (R) or absence (nR) or Arginine residue in position 52 of DQA1 locus, are reported in the abscissa. For each combination of the values of statistical parameters Pc and Relative Risk (RR) are reported in abscissa.

Taken together these results strongly suggest that the presence of homozygous DQB1(nD)-DQA1(R) genotype should be considered as the highest risk marker for susceptibility to IDDM in the population. Screening of population for preventive purposes and/or earliest sign of IDDM might then take advantage of this result of molecular DQ genotyping. "Susceptible homozygous" individuals should be followed very closely and considered the first group of choice for possible new therapeutical trials.

It is still highly debated whether HLA and disease association is directly dependent on the classical polymorphic products of HLA, such as in this case the DQ molecules, or to the products of other unknown or still ill-defined genes in strong linkage disequilibrium with certain HLA-DQ alleles [13]. If, as it seems more likely, susceptibility or resistance to IDDM correlates with the molecules encoded by class II genes, a general consideration should always be present: the specificity of binding a given peptide by a class II molecule is the result of a sufficient affinity and of a comparatively lower amount of other class II molecules with cross reactivity for the same peptide. In this case two possible, but not necessarily opposing, scenarios can be envisaged. In the first case, self antigenic peptides derived by anomalous degradation of pancreas components, for which no immunological tolerance had been acquired during ontogeny, can be bound by DQ α-β heterodimers of certain specificities and presented to autoreactive T cell clones. If the number of possible DQ α-β heterodimers formed is high and of sufficient affinity for a family of peptides, autoreactive T cells can be triggered. This can explain the higher risk of contracting IDDM by DQB1(nD)-DQA1(R) homozygous subjects which can form 4 susceptible heterodimers (2 in cis and 2 in trans). The presence of an Aspartic acid in position 57 of DQB1 alleles would be the major element of protection. The particular position of this residue in the beta chain of the DQ molecule, at one side of the alpha-helix, could permit, for example, the formation of a strong electrostatic interaction with the basic residue Arginine in position 79 of the DQ alpha chain [14], this imposing on the DQ α-β heterodimer a peculiar tridimensional structure at one end of the supposed antigen pocket. The other scenario due to a direct involvement of DQ molecules may relate to the selection of the T cell repertoire in the thymus (Figure 2).

In this case particular DQ molecules, like the ones associated to susceptibility, expressed in antigen presenting cells of the thymus might serve to shape up the repertoire of T cells. If these DQ molecules have relatively low affinity for self peptides analogous to those of pancreatic beta cell proteins involved later in life in the autoimmune attack, specific autoreactive T cell clones can escape for example clonal deletion and be exported in the periphery (Tp2 in Figure 2A). Alternatively, presentation by susceptible DQ molecules of a specific family of self peptides can be highly performant; this may induce preferential deletion of certain T cell clones and by consequence may facilitate the amplification of other, potentially autoreactive T cell clones restricted by HLA class II molecules other than DQ (Tp3 in Figure 2B). In this case susceptibility will be always associated to that particular DQ specificity but autoreactive T cell damage will be mediated by other class II specificities. Further investigation on the biochemical nature of putative pancreas autoantigenic peptides and on the repertoire of peptides that can be bound by appropriate class II molecules (such as "susceptible" or "resistant" DQ α-β heterodimers) will certainly contribute to a better understanding of immunopathogenesis of IDDM.

HLA-DQ and IDDM susceptibility : "Periphery" or "Thymic" dangerous liaisons ?

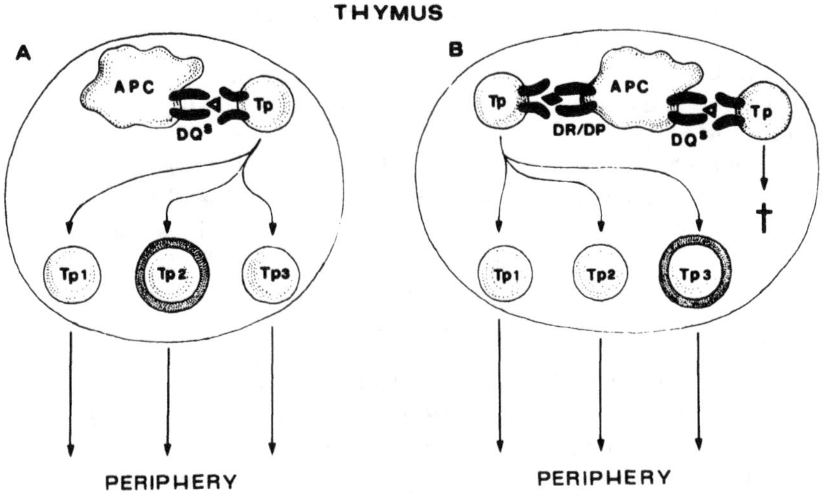

Figure 2. Hypothetical models for intrathymic selection of autoreactive T cell clones influenced by the HLA-DQ individual's genotype. For explanation of the models see the text.

Acknowledgements

This work was supported by the following grants: Conbiotec PNR "Nucleic Acid Probes"; CNR PF Biotechnologies and Bioinstrumentation; CNR PR Genetic Engineering; MURST 40% Neuriommunology; AIDS M.S. and I.S.S. National Research Project; ARBI.

References

1. Foulis AK, Liddle CN, Farquharson MA, Richmond JA, Weir RS. The histopathology of the pancreas in Type I (insulin-dependent) diabetes mellitus: A 25-year review of deaths in patients under 20 years of age in the United Kingdom. Diabetologia 1986;29:267-74.
2. Castano L, Eisenbart GS. Type I diabetes: A chronic autoimmune disease of human, mouse and rat. Annu Rev Immunol 1990;8:647-79.
3. Garchon H-J, Bedossa P, Eloy L, Bach J-F. Identification and mapping to chromosome 1 of a susceptibility locus for periinsulitis in non-obese diabetic

mice. Nature 1991;353:260-62.
4. Cornall RJ, Prins J-B, Todd JA, et al. Type 1 diabetes in mice is linked to the interleukin- 1 receptor and Lsh/Ity/Bcg genes on chromosome 1. Nature 1991;353:262-65.
5. Svejgaard A, Platz P, Ryder LP. HLA and disease 1982- A survey. Immunol Rev 1983;70:193-218.
6. Carra G, Accolla RS. Structural analysis of human Ia antigens reveals the existence of a fourth molecular subset distinct from DP, DQ, DR molecules. J Exp Med 1987;165:47-63.
7. Nepom BS, Palmer J, Kim SJ, Hansen JA, Holbeck SL, Nepom GT. Specific genomic markers of the HLA-DQ subregion discriminate betweeen DR4+ insulin-dependent diabetes mellitus and DR4+ seropositive juvenile rheumatoid arthritis. J Exp Med 1986;164:345-50.
8. Tod JA, Bell JI, McDevitt HO. HLA-DQB gene contributes to susceptibility and resistance to insulin dependent diabetes mellitus. Nature 1987;329:599-604.
9. Morel PA, Dorman JS, Todd JA, McDevitt HO, Trucco M. Aspartic acid at position 57 of the HLA DQB chain protects against type I diabetes: A family study. Proc Natl Acad Sci USA 1988;85:8111-115.
10. Nepom GT. HLA and type I diabetes. Immunol Today 1990;11:314-15.
11. Tosi G, Mantero G, Magalini AR, et al. HLA-DQB1 typing of north east Italian IDDM patients using amplified DNA, oligonucleotide probes and a rapid DNA-enzyme immunoassay (DEIA). Mol Immunol 1993;30:69-76.
12. Khalil I, d'Auriol L, Gobet M, et al. A combination of HLA DQB Asp 57-negative and HLA DQA Arg 52 confers susceptibility to insulin-dependent diabetes mellitus. J Clin Invest 1990;85:1315-319.
13. Altmann DM, Sansom D, Marsh SGE. What is the basis for HLA-DQ associations with autoimmune disease? Immunol Today 1991;12:267-70.
14. Marsh SGE, Bodmer JG. HLA class II nucleotide sequences, 1992. Immunogenetics 1993;37:79-94.

NEW CLINICAL AND MOLECULAR FEATURES OF "ESSENTIAL" MIXED CRYOGLOBULINEMIA

Domenico Sansonno, Vito Cornacchiulo, Anna Rina Iacobelli, and Franco Dammacco
Department of Biomedical Sciences and Human Oncology
Section of Internal Medicine and Clinical Oncology
University of Bari Medical School
Bari
ITALY

Introduction

Cryoglobulinemia, first described by Lerner and Watson in 1947 [1], is characterized histologically by a small vessel vasculitis which appears to be immune-complex mediated. Three kinds of cryoglobulins are recognized: a) Type I cryoglobulins are single monoclonal Ig found in malignant disorders such as Waldenström's macroglobulinemia and multiple myeloma; b) Type II cryoglobulins are monoclonal Ig, usually IgMk with rheumatoid factor (RF) activity, complexed with polyclonal IgG; c) Type III cryoglobulins are polyclonal Ig, usually IgG and IgM but sometimes IgA, one of which has RF activity.

Types II and III are uncommon in Northern Europe and North America and are frequently associated with chronic active liver disease (CALD) [2]. Serological evidence of chronic infection with hepatitis B virus (HBV) in some patients was initially seen as proof of a viral involvement in cryoglobulinemia [3], though subsequent work showed that such infection was uncommon [4]. An association with hepatitis A virus infection was described in the middle 1980s [5]. The question has subsequently been extensively reexamined since the introduction [6] of diagnostic testing for hepatitis C virus (HCV).

Mixed Cryoglobulinemia (MC) and HCV Infection

Initial reports drew attention to the high prevalence of anti-HCV antibodies in MC [7-10], and an association between the two disorders appeared more than coincidental even after elimination of confounding factor of nonspecific anti-HCV seropositivity.

Detection of circulating anti-HVC antibodies should be regarded as provisional

A. G. Dalgleish et al. (eds.), The Impact of Biotechnology on Autoimmunity, 27–36.

when "sticky" serum samples are tested, since false-positive reactions have been obtained with first-generation ELISA in sera from patients with monoclonal components, autoimmune chronic active hepatitis, as well as rheumatoid arthritis [11-13].

The second-generation HCV ELISA (c200/c22-3) has improved both sensitivity in individuals at high risk for HCV infection, and specificity when screening low risk populations. Cryoglobulinemic sera are "sticky" samples. We have none-the-less shown that this assay is a reliable test. It demonstrated good repeatibility in more than 90% of the samples and was confirmed by immunoblotting (RIBA). Neither the cryocrit levels nor the titres of RF activity influenced the results [10].

We have studied 62 MC patients subdivided as follows: 35 with cryoglobulinemia associated with the purpura-weakness-arthralgia syndrome without evidence of liver involvement (group 1); 15 with symptomatic MC and CALD (group 2); 12 with CALD and without MC (group 3). Anti-HCV antibodies were found in all three groups: 24/35 (68.6%), 15/15, and 12/12, respectively. The association was rendered more striking by the detection of HCV genomic sequences in purified cryoglobulins by reverse transcription and cDNA amplification, using a polymerase chain reaction (PCR) [14-16]. The demonstration of HCV RNA in the cryoprecipitates (fig. 1) suggested that cold-insoluble immune-complexes include intact virions bound to IgG specific antibodies.

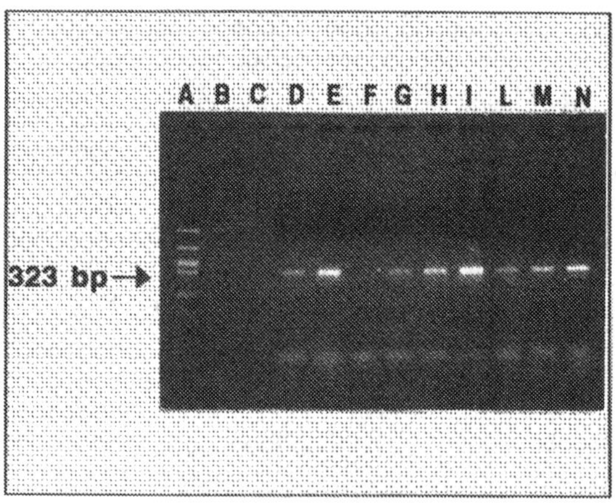

Fig. 1: Hepatitis C virus RNA detection by PCR, in each patients amplifying the 5'NC region of the genome. DNA products were analyzed by agarose gel electrophoresis. Lane A, pGem molecular weight marker; lane B, negative control; lane C-N, cryoglobulinemic samples; bp, base pairs.

HCV RNA was also detected in 6 of 11 anti-HCV negative patients in group 1, indicating that current serological tests underestimate HCV infection and provide indirect evidence of unidentified anti-HCV antibodies directed against antigenic determinants probably expressed on intact virions [16].

No association with HCV infection was found in patients with type I (IgMk) cryoglobulinemia and a clinical diagnosis of Waldenström's macroglobulinemia. The mechanisms responsible for type I cryoglobulinemia may therefore be entirely different from those underlying type II and possibly type III cryoglobulin formation [16].

Anti-GOR Antibodies: Specific HCV-Induced Autoimmunity

Genetic analysis in experimental models indicates that cryoglobulin production is crucial for the development of autoimmune phenomena including necrotizing vasculitis and glomerulonephritis. Increasing evidence suggests that peculiar genetic expression in B cells is essential for autoantibody production, although still dependent on T-cell help. Monoclonal antibodies with RF activity induce cutaneous vasculitis, as well as glomerulonephritis, whereas antibodies lacking this activity seem only capable of inducing glomerulonephritis [17]. Therefore, it is essential to establish whether cryoglobulin production associated with HCV infection is an autonomous self-perpetuating process due to an autoimmune response against a virus-induced autoantigen, or part of the expression of a peculiar clinical entity subsequent to chronic viral stimulation.

Mishiro et al. [18] have described antibodies against GOR antigen, a fusion protein expressed by a cDNA clone (GOR 47-1) from a chimpanzee infected with hepatitis C virus. Anti-GOR antibodies appear to be strikingly associated with HCV infection, and are said to reflect specific HCV-induced autoimmunity. To investigate their role in MC associated with HCV infection, we searched for anti-GOR immune response in 44 patients with ongoing HCV infection with and without MC. HCV RNA and specific antibody activities of IgG and IgM components of purified cryoprecipitates were tested. Patients were divided as described in the previous section.

Table 1. Anti-HCV antibodies, HCV RNA, and anti-GOR antibodies in patients with and without mixed cryoglobulinemia.

	PATIENTS	ANTI-HCV (%) ELISA	RIBA	HCV RNA (%)	ANTI-GOR (%)
GROUP 1 (13)	SUPERNATANTS	8 (61.5)	9 (69.2)	11 (84.6)	9 (69.2)
	CRYOPRECIPITATES	9 (69.2)	9 (69.2)	11 (84.6)	9 (69.2)
GROUP 2 (12)	SUPERNATANTS	9 (75)	12 (100)	11 (91.7)	9 (75)
	CRYOPRECIPITATES	10	11	11 (91.7)	6 (50)
GROUP 3 (19)	SERUM	19 (100)	19 (100)	19 (100)	6 (31.6)

Anti-GOR antibodies were found in 9/13 (69.2%) supernatants from group 1, and in 9/12 (75%) of group 2 (table 1). Those in the corresponding cryoprecipitates identified another two reactive samples from group 1 that were anti-HCV negative but HCV-correlated, as demonstrated by the presence of HCV RNA.

Purified and fractionated cryoprecipitates revealed further anti-GOR reactivity, suggesting that serological studies may sometimes result in false-negatives. The overall prevalence of anti-GOR antibodies in patients with MC (group 1 plus group 2) was 72% (18/25), significantly ($p<0.05$) different from the 31.6% (6/19) in group 3. All anti-GOR positive patients were also positive for HCV RNA.

As shown in fig. 2, no anti-GOR activity was found in the IgM peak, whereas it was present in the IgG peak, indicating that anti-GOR reactivity is not due to RF. Thus, it appears that IgM-RF may be bound to anti-GOR IgG in the cryoprecipitating IC.

Our previous observations indicated that MC is found asymptomatically in 13% of chronically HCV-infected patients, whose cryoprecipitates are said to contain concentrated IC, presumably formed by complete HCV virions, with the corresponding antibodies cross-linked by RF mostly belonging to the WA group. HCV is thus a likely cause of cryoprecipitation, though since this only occurs in a minority of HCV-infected patients, other mechanisms may be involved.

The association between MC and anti-GOR immune response supports the view that anti-GOR autoimmunity may play a role in the production of cryoglobulins. GOR antigen is the encoded product of a single-copy gene of host cellular sequence induced by HCV infection, and seems to reflect specific HCV-related autoimmunity. The HCV-restricted nature of anti-GOR autoantibodies seems to be the result of molecular mimicry between GOR, a nuclear host-derived antigen, and the core gene product of HCV [19].

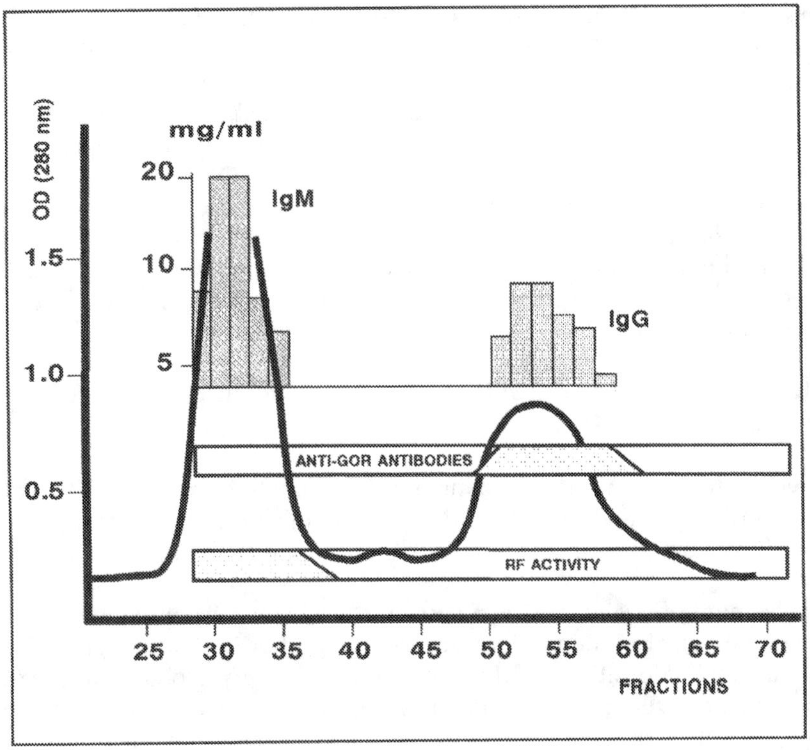

Figure 2. Chromatographic separation of IgM and IgG from a purified cryoprecipitate fractionated on Ultrogel AcA 34 column. The solid line represents the protein elution profile (optical density 280 nm). The inserted diagrams demonstrate IgM and IgG reactivities analyzed by nephelometry. Anti-GOR antibodies and RF activity have been demonstrated in the peaks corresponding to the dotted regions.

The function and biological significance of GOR are unknown. Observation of anti-GOR antibodies permits better differentiation of type-2 chronic autoimmune hepatitis [20]. The presence of anti-GOR autoantibodies in the cryoprecipitates points to their involvement in the extrahepatic manifestations of MC. Johnson *et al.* [21] have stressed HCV infection as a cause of membranoproliferative glomerulonephritis. They demonstrated the subendothelial deposition of IC, but were unable to identify HCV antigens or HCV RNA within diseased glomeruli.

Thus, immune deposits could be different in nature and related to other types of IC. This is consistent with our results, in that anti-GOR antibodies can circulate in the absence of anti-HCV antibodies in HCV-chronically infected patients, as demonstrated in the 2 anti-HCV negative, HCV RNA-positive patients whose cryoglobulins contained anti-GOR autoantibodies. Other antibodies directed to unrecognized HCV structural surface antigens or other viral epitopes associated with intact virions cannot be ruled out. Even so, anti-GOR autoantibodies may be involved in the pathogenesis of cold-precipitable IC and of the tissue damage at precipitation sites.

MC may be regarded as an autoimmune disease superimposed on chronic HCV infection. This immune response to HCV-induced autoantigen(s) may proceed to an autonomous autoimmune disease in predisposed patients.

Identification of HCV-Associated Proteins in Skin and Muscle Tissues of Anti-HCV Positive Patients with MC

Cryoprecipitates in HCV-infected patients are essentially composed of IC containing virions specifically bound to IgG anti-HCV antibodies cross-linked by IgM with RF activity. HCV genomic sequences are sometimes found even in anti-HCV negative patients.

Involvement of cryoglobulins in vasculitis has been substantiated clinically and experimentally, and by immunofluorescence studies showing immunoglobulins in the vessel wall. IC deposition is thought to underlie cryoglobulinemic vasculitis. Circulating IC are generally deposited in vessel walls. Fixation of complement then results in their necrosis and deposition of fibrin. Deposits of immunoglobulins of the same classes as those demonstrable in the cryoglobulins in both skin lesions and the diseased glomeruli of patients with kidney involvement, have formed the subject of several studies. Specific demonstration of the antigen involved, however, has received little attention [22].

Demonstration of HCV deposition at the site of the extrahepatic manifestations in MC patients is essential to show its direct involvement in such damage. HCV-associated antigens by immune detection, however, cannot readily be demonstrated nor HCV RNA specific viral sequences by *in situ* hybridization in tissues.

We have shown that tissue expression of c-100-3 antigen, a NS4-encoded protein, is labile and sensitive to the use of fixatives during tissue processing [23]. As HCV circulates in autoantibody excess, antigenic determinants may be masked by specific antibodies [15]. Our preliminary studies on the composition of circulating IC in anti-HCV positive patients demonstrated that HCV induces the production of IgM RF, which is an essential component of a virus-induced autoimmune response. It was thus clear that the virus regularly circulates in a complexed form, since specific viral determinants are bound to IgG anti-HCV antibodies, which in turn are cross-linked by IgM RFs.

We then unmasked specific HCV epitopes in tissues processed for immunohistochemistry by treating them with electrophoresis under acid conditions (acid electroelution). This efficiently elutes Ig components from masked antigens without lowering or spoiling the antigenicity of HCV-associated proteins (HCV-Ag). HCV-Ag deposition was demonstrated in vascular structures of skin and muscle tissues in 5 (33.9%) and in 2 (13.3%) respectively of 15 MC patients plus HCV infection (fig. 3).

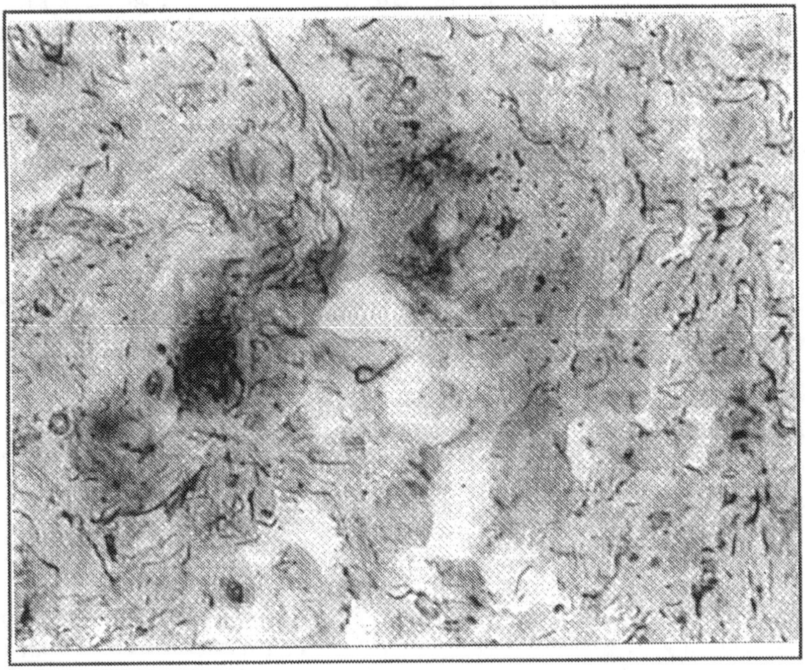

Figure 3. HCV-Ag immune deposits within dermal vessel walls and perivascular spaces of an anti-HCV-positive cryoglobulinemic patient.

Two types of immunohistochemical pictures were observed: one with coarse intravascular deposits of HCV-Ag that sometimes completely fill the vessel lumen and are associated with dermal inflammatory infiltration and vasculitis; the other with deposition confined to the vessel walls in histologically normal tissue without vasculitis. Wall deposition preferentially involves small arteries, and is found within and between endothelial cells up to the basal lamina. This could reflect the mechanism by which HCV-Ag deposits in the wall come from the serum up to the basal membrane, and positivity in the endothelium may be the result of the phagocytosis demonstrated in both damaged and undamaged vessels in MC. Perivascular immune deposits and inflammation may be regarded as the result of local complement fixation, with subsequent necrosis of the vessel walls and fibrin deposition. No correlation has been found between HCV-Ag vascular deposition in the skin and clinical and laboratory parameters, including cryocrit values, HCV viremia, anti-HCV immune response, clinical manifestations of cutaneous vasculitis, type and severity of liver damage, and type of treatment or its lack.

These data reiterate the concept that the cryoprecipitates are intrinsically capable of damaging tissues. This property may be related to the biological activities of IC. Our results indicate that tissue deposition of HCV-Ag is not enough to cause active vasculitis.

The apparent lack of HCV-Ag deposits in almost 70% of our series may indicate that IC are different in nature in these cases, as suggested elsewhere [20]. However, this could equally be attributable to the limits of our immunohistochemical method. More potent antibodies or more sensitive techniques such as *in situ* polymerase chain reaction, may show whether or not such a difference exists.

Acknowledgements

The excellent secretarial assistance of Gianfranco Lauletta and Vito Iacovizzi is gratefully acknowledged. This study was supported by grants from the Italian Ministry for Universities and Scientific and Technological Research, group "Liver Cirrhosis", Rome, and local funds 60%

References

1. Lerner AB, Watson CJ. Studies of cryoglobulins. I. Unusual purpura associated with the presence of a high concentration of cryoglobulin (cold precipitable serum globulin). Am J Med Sci 1947; 214: 410-415.
2. Brouet JC, Clauvel JP, Danon F, Klein M, Seligmann M. Biologic and clinical significance of cryoglobulins: a report of 86 cases. Am J Med 1974; 57: 775-778.
3. Levo Y, Goreic PD, Kassab HJ, Zucker-Franklin D, Franklin EC. Association between hepatitis B virus and essential mixed cryoglobulinemia. N Engl J Med 1977; 246: 1501-1504.
4. Popp JW, Dienstag JL, Wands JR, Block KJ. Essential mixed cryoglobulinemia without evidence for hepatitis B virus infections. Ann Intern Med 1980; 92: 379-383.
5. Innman RD, Hodge M, Johnston MEA, Wright J, Heathcote J. Arthritis,

vasculitis and cryoglobulinemia associated with relapsing hepatitis A infection. Ann Intern Med 1986; 105: 700-703.

6. Choo QL, Kuo G, Weiner AJ, Overby LR, Bradley DW, Hougthon M. Isolation of cDNA clone derived from a blood borne non-A, non-B viral hepatitis genome. Science 1989; 244: 359-362.
7. Pascual M, Perrin M, Giostra E, Schifferli JA. Hepatitis C virus in patients with cryoglobulinaemia tipe II. J Infect Dis 1990; 162: 569-570.
8. Disdier P, Harl. JR, Weiller PJ. Cryoglobulinaemia and hepatitis C infection. Lancet 1991; 338: 1151-1152.
9. Ferri C, Greco F, Longobardo G, et al. Antibodies to hepatitis C virus in patients with mixed cryoglobulinemia. Arthritis Rheum 1991; 34: 1606-1610.
10. Dammacco F, Sansonno D. Antibodies to hepatitis C virus in essential mixed cryoglobulinaemia. Clin Exp Immunol 1992; 87: 352-356.
11. Mc Farlane IG, Smith HM, Johnson PJ, Bray GP, Vergani D, Williams R. Hepatitis C virus antibodies in chronic active hepatitis; pathogenetic factor or false-positive result? Lancet 1990; 335: 754-757.
12. Theilmann L, Blazek M, Goeser T, Gmelin K, Kommerel B, Fiehn W. False positive anti-HCV tests in rheumatoid arthritis. Lancet 1990; 335: 1346.
13. Boudart D, Lucas JG, Muller JV, Lecarrer D, Planchon B, Harousseau JL. False positive hepatitis C virus antibody tests in paraproteinemia. Lancet 1990; 336: 63.
14. Pechere-Bertschi A, Perrin L, De Saussure P, Widmann JJ, Giostra E, Schifferli JA. Hepatitis C: a possible etiology for cryoglobulinaemia type II. Clin Exp Immunol 1992; 89: 419-422.
15. Agnello V, Chung RT, Kaplan LM. A role for hepatitis C virus infection in type II cryoglobulinemia. N Engl J Med 1992; 327: 1490-1495.
16. Dammacco F, Sansonno D, Cornacchiulo V, et al. Hepatitis C virus infection and mixed cryoglobulinemia: a striking association. Int J Clin Lab Res 1993; 23: 45-49.
17. Denman AM. Cryoglobulins and immunopathological manifestations of autoimmune disease. Clin Exp Immunol 1992; 87: 169-171.
18. Mishiro S, Hoshi J, Takeda K, et al. Non-A, non-B hepatitis specific antibodies directed at host-derived epitope: implication for an autoimmune process. Lancet 1990; 338: 1400-1403.
19. Mishiro S, Takeda K, Hoshi Y, Yoshikawa A, Gotanda T, Itoh Y. An autoantibody cross-reactive to hepatitis C virus core and a host nuclear antigen. Autoimmunity 1991; 10: 269-273.
20. Michel G, Ritter A, Gerken G, Meyer zum Buschenfelde KH, Decker R, Manns MP. Anti-GOR and hepatitis C virus in autoimmune liver disease. Lancet 1992; 339: 267-269.
21. Johnson RJ, Gretch DR, Yamabe H, et al. Membranoproliferative glomerulonephritis associated with hepatitis C virus infection. N Engl J Med 1993; 328: 465-570.

22. Farivar M, Wans JR, Benson GD, Dienstag JL, Isselbacher KJ. Cryoprotein complexes and peripheral neuropathy in a patient with chronic active hepatitis. Gastroenterology 1976; 71: 490-493.
23. Sansonno D, Dammacco F. Hepatitis C virus c100 antigen in liver tissue from patients with acute and chronic infection. Hepatology (in press).

POST-STREPTOCOCCAL AUTOIMMUNE SEQUELAE: A LINK BETWEEN INFECTION AND AUTOIMMUNITY

Malak Kotb, Ph.D.
University of Tennessee, Memphis
VA Medical Center (151)
1030 Jefferson Avenue
Memphis, TN 38104
USA

Introduction

For many years microbial pathogens have been implicated in the development of autoimmunity [1-8]. Several studies have attempted to link viral infections with rheumatoid arthritis (RA), insulin-dependent diabetes (IDDM), myasthenia gravis, systemic lupus erythematosus (SLE), myocarditis and AIDS [2,8-15]. Candidate viruses included Coxsackie virus, Epstein-Barr virus, Herpes simplex virus, parvovirus, Type C retroviruses, vesicular stomatitis virus and rubella virus [8,3]. Similarly, mycoplasmas have been linked to RA [16], mycobacteria to RA and SLE [17,18], and infection with *klebsiella sp* has been implicated in the pathogenesis of ankylosing spondylitis [19]. In some of these instances the association between infection and autoimmune disease fall short of a definitive proof; however, there are numerous examples where it is clear that microbial infections result in autoimmune manifestations. Infection with HIV can cause psoriasis, Reiter's syndrome and induce autoantibody generation [20,21]. Lyme disease is associated with Borreliosis [22] and Reiter's syndrome has been reported to develop following Gram-negative enteritis and nongonococcal urethritis [6,23]. Of particular interest to us are group A streptococcal (GAS) infections which can trigger acute rheumatic fever (ARF), rheumatic heart disease (RHD) and glomerulonephritis in susceptible hosts.

How can infectious pathogens induce autoimmunity? Several hypotheses have been proposed; of these three seem most attractive. The first hypothesis is that autoimmunity may be a consequence of molecular mimicry between microbial and host proteins. The response to a dominant microbial protein epitope may evoke an immune response against a cross-reactive cryptic autoimmune epitope thereby inducing autoreactive T cells and eliciting autoimmune antibodies both of which can cause damage to host tissues. Alternatively, due to molecular mimicry, the pathogen may escape immune destruction and persist in the host causing tissue damage and disease. A second hypothesis is that pathogens can produce molecules that are

A. G. Dalgleish et al. (eds.), The Impact of Biotechnology on Autoimmunity, 37–50.

potent inducers of inflammatory cytokines such as interferon-γ (IFN-γ) which can upregulate the expression of MHC molecules and induce the expression of class II molecules on cells that normally express little or no class II [24-26]. This can result in presentation of sequestered self antigens and the initiation of autoimmunity. The third and most recent hypothesis proposes that autoimmunity may be triggered by superantigens which are produced by certain pathogens and which are potent inducers of T cell proliferation and inflammatory cytokines [27,28]. Unlike conventional antigens, superantigens react with less variant regions within the variable portion of the β chain of the T cell receptor αβ heterodimer. These regions are known as Vβ elements, and there are 24 major TCR Vβ families in humans. Each superantigen has a characteristic affinity to a set of Vβ elements and stimulates all T cells that express those elements without regard to the antigenic specificity of each TCR. This kind of stimulation can theoretically induce the expansion of autoreactive T cell clones that have escaped elimination in the thymus [27].

It is clear that none of the above hypotheses are mutually exclusive, and in fact, they most likely act in concert to initiate autoimmune disease. Although events leading to the onset of most autoimmune diseases remain obscure, the development of new technologies, in particular PCR has greatly impacted on our knowledge of the pathogenesis of these diseases. Reports of restricted TCR Vβ usage in a number of autoimmune disorders including RA, MS, and Sjogren's syndrome [29-34] have suggested a role for superantigens in disease pathogenesis and provided a strong link between infection and autoimmunity. The concept that superantigens act by expanding autoreactive T cell clones bearing superantigen-specific Vβ elements is extremely appealing; however, direct proof for this hypothesis is yet to be obtained.

Poststreptococcal autoimmune diseases such as ARF and RHD provide excellent models for studying the infectious etiology of autoreactivity and can be used to decipher the role of superantigens in disease. Rheumatogenic M protein serotypes (12 out of 80 serotypes) are highly associated with ARF [35,36]. Historically ARF and RHD were considered diseases of molecular mimicry, mediated by autoreactive antibodies elicited by M protein epitopes that mimic host proteins [37-39]. However, like most organ-specific autoimmune diseases, there is strong evidence that T cells have a major role in ARF and RHD. We have shown that certain strains of *S. pyogenes* produce a number of superantigens including the pyrogenic exotoxins and the M proteins [40-42]. This discovery suggests pathways by which activated T cells can induce disease pathogenesis. Studies with pepsin extracted M proteins revealed that each serotype has a unique pattern of human Vβ specificity, but all rheumatogenic serotypes tested share the ability to stimulate Vβ4 T cells. To determine whether stimulation of Vβ4 cells may be related to the a-helical coiled-coil conformation of the M protein, we asked whether a protein such as myosin which is structurally similar to the M proteins can also act as a superantigen. Although myosin failed to induce T cell blastogenesis, it did stimulate the proliferation of M protein-primed T cells, but not anti-CD3 primed T cells. These data provide a link between the superantigen and the molecular mimicry hypothesis and suggest that recognition of cardiac proteins by T cells can occur following induction by streptococcal superantigens. This may also account for the target organ specificity of these diseases. However, we believe that neither superantigenic stimulation by the M proteins or other streptococcal superantigens nor the presence of crossreactive epitopes can alone be responsible for the initiation of the disease. Environmental and genetic factors affecting susceptibility are likely to interact with the pathogen to create disease. Understanding these interactions will undoubtedly enhance our ability to detect susceptible individuals and treat these diseases.

Materials and Methods

ISOLATION AND PURIFICATION OF M PROTEIN

M proteins were purified by limited pepsin digestion of types 5, 6, 18, 19 and 24 group A streptococci as previously described [43] and referred to herein as pep M proteins. Purified pep M serotypes 6, 18, 19, and 24 were provided to us by the late Dr. Edwin H. Beachey. Pepsin extracted pep M5 was purified in our laboratory by HPLC DEAE (Protein PAK 5 PW) anion exchange chromatography [43]. Recombinant pep M5 was generated and purified as described elsewhere [44].

ISOLATION AND STIMULATION OF HUMAN LYMPHOCYTES.

Peripheral blood mononuclear cells (PBMC) were isolated from heparinized on Ficoll-Hypaque gradients. PBMC were further purified by 1 cycle of erythrocyte (E)-rosetting to enrich for T cells. T cells were cultured at 10^6 cells/ml in RPMI-1640 medium containing 10% heat-inactivated fetal bovine serum and supplemented with 4 mM L-glutamine, 50 U/ml penicillin G, and 50 μg/ml streptomycin (RPMI-complete medium). Unless otherwise indicated T cell cultures were stimulated with 2 μg/ml anti-CD3 mAb, OKT3 (anti-CD3, IgG_{2a}) or 1 μg/ml of the various pep M preparations. The cultures were incubated for 3 days at 37°C in a humidified atmosphere of 5% CO_2.

Highly purified T cells were obtained by subjecting freshly isolated PBMC to 2 cycles of E-rosetting followed by overnight adherence at 37°C in RPMI-complete medium. The E-rosette negative population enriched in B cells and monocytes were used as antigen presenting cells (APC).

MYOSIN PREPARATION

Canadian Albino rabbit heart myosin was purified as described [45]. The purity of the myosin preparation was verified by comassie blue-stained SDS-PAGE.

ANALYSIS OF TCR Vβ GENE USAGE BY PCR

Analysis of Vβ usage was performed according to a slight modification [41] of the method described by Choi et al. [46]. Total RNA extracted from each culture was reverse transcribed into cDNA. The cDNA from each culture was aliquoted into 22 tubes, each containing a 5' Vβ-specific primer representing the different Vβ families along with a 3' Cβ anti-sense primer amplifying products ranging from 170-220 base pairs in length. As an internal control, each tube contained 5' and 3' Cα primers which amplified a 600 base pair product. All primers were used at a final concentration of 0.3 μM. Amplification for TCR Vβ usage was performed as previously described [41,42]. Quantitation of the amplified products was achieved by amplifying the cDNA in the presence of 3' ^{32}P-labeled Cα and Cβ primers (10^6 cpm of each per reaction). Radiolabeled PCR products were separated on 2% agarose gels and exposed to x-ray film. The relative amounts of Cα and Vβ bands were determined by scanning the autoradiograms and integrating the areas of the relevant peaks. PCR values were normalized by dividing the area of each Vβ by their respective Cα internal control. The normalized values for pep M-stimulated cells were divided by those obtained for anti-CD3-stimulated cells to determine specific expansion of Vβ elements. The PCR value = $[\text{area } V\beta_n / \text{ area } C\alpha]_{\text{pep M}}$ divided by $[\text{area } V\beta_n / \text{ area } C\alpha]_{\text{anti-CD3}}$, where n= 1-20.

Results

SUPERANTIGENICITY OF M PROTEIN

To be classified as a superantigen, a protein has to meet certain criteria [28]. First, it has to evoke powerful T cell blastogenesis; second, its ability to stimulate T cells is dependent on the presence of MHC class II-expressing antigen presenting cells (APC); third, the presentation of the protein to T cells is MHC-nonrestricted; fourth, it can bind directly to HLA molecules without need for processing by APC. The fifth, and most important criteria is that the protein can induce the preferential expansion of a limited set of Vβ bearing T cells and that the expanded T cells show no evidence of oligoclonality. The characteristic affinity of a superantigen to a particular set of Vβ elements does not vary from one individual to another.

Earlier studies from our laboratory [47] revealed that pep M5 is a potent stimulator of human T cells and that it can be presented by allogeneic APC [48]. Unlike the response of T cells to polyclonal mitogens such as PHA which can be supported by class II negative accessory cells, the response to pep M5 required that these cells express class II molecules on their surface. This was confirmed by studies in which mouse T cells transfected with human HLA DR, DQ or DP elements were capable of presenting pep M5 to human T cells; whereas, fibroblasts transfected with vector lacking HLA genes were unable to support the pep M5 response but were able to support the response to PHA (Figure 1). These data show that the response to pep M5 requires class II molecules for its ability to stimulate T cells, but that its presentation to T cells is MHC-unrestricted and independent of the class II isotype or haplotype. In addition, further studies revealed that processing of pep M5 by the APC was not required for its ability to stimulate T cells, and that pep M5 can bind directly to HLA molecules [40].

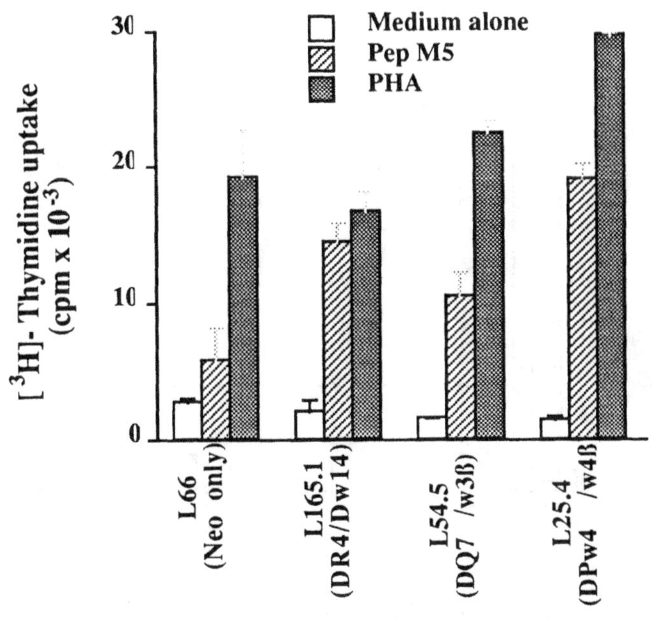

Figure 1: T cells (10^5) were cultured with or without 2 X 10^4 mitomycin C -treated mouse fibroblasts transfected with: neomycin resistance gene only (L66), HLA-DR (L165.1), HLA-DQ (L54.5), or HLA-DP(L25.4). Cultures were incubated for 3 days in medium alone or in the presence of pep M5 or PHA (5μg/ml). Proliferation was assessed by [^{3}H]-thymidine incorporation. The data are expressed as the mean cpm of triplicate cultures ±SEM. Data adapted from Tomai et al, 1990 J. Exp. Med. [40].

To obtain direct evidence for the superantigenicity of the pep M5 protein, we examined its ability to stimulate T cells in a Vβ specific manner. To examine the expansion of specific Vβ families one can either use monoclonal anti-Vβ-specific antibodies and flow cytometry or analyze the expression of Vβ-specific mRNA by PCR. Due to the limited number of anti-human Vβ antibodies available, the highly sensitive PCR technique was used. Parallel cultures of T cells were stimulated with either superantigen or with anti-CD3 antibody which is a polyclonal T cell mitogen. Anti-CD3 stimulation results in proportional amplification of all TCR Vβ genes so that the relative expression of the various Vβ families in the anti-CD3 stimulated populations reflects the normal baseline repertoire of each individual. Therefore, if analysis of the Vβ usage by superantigen-stimulated T cells shows expansion of particular Vβ families over and above that which is seen in the anti-CD3-stimulated cells this indicates that these Vβs are preferentially used in response to this superantigen. As shown in Figure 2, stimulation by pep M5 resulted in specific expansion of Vβ2, Vβ4, and Vβ8-bearing T cells [41,42].

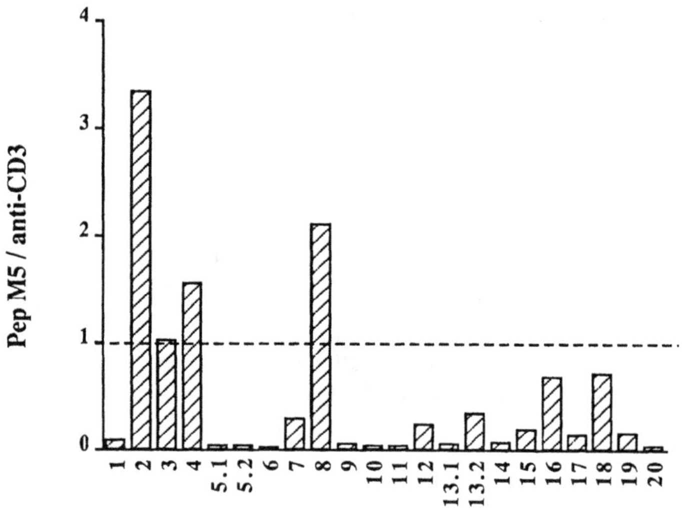

Figure 2. PCR analysis of the expression of the entire panel of TCR Vβ families in pep M5-stimulated T cells. Purified T cells (10^6) from an individual were stimulated for 3 days in the presence of irradiated APC with either anti-CD3 (10μg) or pep M5 (1 μg/ml). RNA was extracted from the cells, cDNA prepared and amplified by PCR. The radioactive PCR products were separated on agarose gels which were then subjected to autoradiography. Autoradiograms were scanned to obtain a quantitative measure for the amplified Vβ and Cα products. The PCR value = [area $V\beta_n$/ area $C\alpha$]$_{\text{pep M}}$ divided by [area $V\beta_n$ / area $C\alpha$]$_{\text{anti-CD3}}$, where n= 1-20.

To eliminate the possibility that pep M5 may be contaminated with streptococcal pyrogenic exotoxins, we compared the Vβ specificity of these toxins with that of pep M5. As shown in Table 1, the Vβ specificity of pep M5 is quite different from that of SPE A, SPE B , or SPE C [42]. Since we did not see expansion of SPE-specific Vβs, this eliminates the possibility of SPE contamination. To rule out contamination by a yet unidentified streptococcal superantigen, we cloned and expressed pep M5 in *E.coli*. The purified recombinant pep M5 stimulated T cells and showed identical Vβ specificity to the native pep M5 [44] thereby confirming that pep M5 is a superantigen.

To determine whether the superantigenicity of pep M5 was unique to this serotype or is a property shared by other pep M proteins, we tested the Vβ specificity of pep M from serotypes that have a high degree of association with rheumatic fever such as serotypes 6, 18, 19 and 24. Our experiments revealed that all of these pep M proteins are superantigens and that although each has a unique pattern of Vβ specificity, they all seem to share an affinity for Vβ4 bearing T cells (Table 1). The ability to stimulate Vβ4 bearing T cells does not appear to be a property of all M proteins, inasmuch as a nonrheumatogenic serotype, pep M2 was found to stimulate only Vβ2 bearing T cells [49].

Table 1. Human TCR Vβ specificity of rheumatogenic streptococcal pep M proteins and pyrogenic exotoxins.

Strep Superantigen	*Human TCR Vβ specificity*							
Pep M 5		Vβ2,	Vβ4,			Vβ8		
Pep M 6	Vβ1,	Vβ2,	Vβ4,			Vβ8		
Pep M 18	Vβ1,	Vβ2,	Vβ4,		Vβ5.2			
Pep M 19			Vβ4,		Vβ5.2,	Vβ8		
Pep M 24	Vβ1,		Vβ4,		Vβ5.2	Vβ8		
SPEA (18)		Vβ2,						Vβ12,Vβ14,Vβ15
SPEB (18)						Vβ8		
SPEC (18)	Vβ1,	Vβ2,		Vβ5.1			Vβ10	

We next asked if other alpha-helical coiled-coil proteins would behave like the M proteins in their ability to act as superantigens to human T cells. We selected myosin because it bears structural homology to M protein, and it has been reported to be a target autoantigen in rheumatic heart disease [50-52]. Purified cardiac myosin failed to stimulate resting human T cells to proliferate suggesting that not all alpha-helical coiled-coil proteins act as superantigens. However, when T cells were first primed with pep M5, they recognized myosin epitopes and proliferated in response to this protein. Figure 3 shows the proliferative response of resting T cells which were first primed for three days with either pep M5 or anti-CD3, then rested for another three days, and finally cultured with myosin for four more days. A significant proliferative response was obtained only with pep M5-primed T cells (Figure 3). Therefore, priming with pep M5 specifically enhances the response to myosin, possibly by enriching for myosin responsive T cells. In addition, our data show that although myosin itself is not a superantigen, it is capable of stimulating T cells expanded by this streptococcal superantigen. This information raises exciting possibilities regarding the role of the M protein in the pathogenesis of rheumatic heart disease [49].

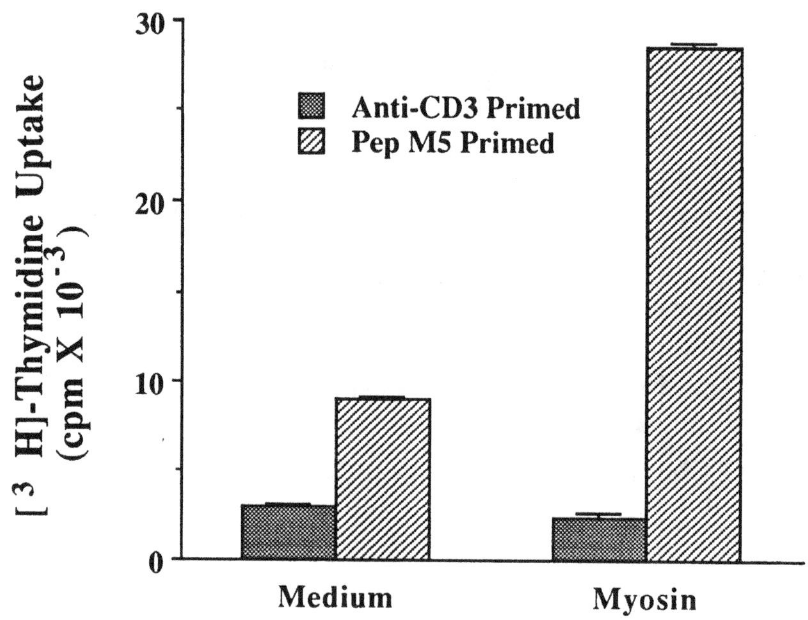

Figure 3. Pep M5-primed human T cells respond to cardiac myosin. T cells (2X10^6/ml) were stimulated with either 1 μg/ml anti-CD3 mAb or 0.5 μg/ml of pep M5 protein. After three days in culture, the cells were washed and placed in fresh RPMI-complete medium for 48 hours to rest. Recovered cells (0.6X 10^5) were then incubated in 96-well plates with either medium alone or with 0.2 μg purified cardiac myosin in a final volume of 200 μl. Proliferation was measured 3 days later by determining [^{3}H]-thymidine uptake. The data are presented as mean cpm of triplicate cultures ± SEM. Data from [49].

Discussion

Post-streptococcal autoimmune diseases offer excellent models for investigating the link between infection and autoimmunity. For years products of certain strains of *S. pyogenes*, in particular the M proteins, have been implicated in the pathogenesis of rheumatic fever and rheumatic heart disease [35]. There are over 80 different serotypes of M protein, but not all are associated with disease. Serotypes that show a high degree of association with acute glomerulonephritis are referred to as nephritogenic serotypes; likewise, serotypes associated with rheumatic fever are referred to as rheumatogenic serotypes [36,53]. There are about eleven rheumatogenic serotypes and these include serotypes M1, M3, M5, M6, M18, M19 and M24. The current studies showing that several rheumatogenic serotypes of M protein are superantigens to human T cells provides a plausible mechanism for their ability to trigger an autoimmune response. Each of the pep M proteins studied has a unique pattern of Vβ specificity and these are distinct from the pattern of Vβ specificity seen with the streptococcal pyrogenic toxins [42,49]. The expansion of many T cell clones by superantigens may result in the stimulation of autoreactive T cells that normally would have been deleted in the thymus during the selection processes. Conceivably, these autoreactive T cells can trigger disease in the affected host [27].

The superantigenicity of the M proteins, or other streptococcal superantigens for that matter, cannot not fully explain the pathogenesis of post-streptococcal autoimmune sequelae. T cells from all individuals tested respond vigorously, and with the same Vβ specificity, to the M proteins; however, only 3% of infected individuals develop post-streptococcal autoimmunity suggesting that there are other contributing factors that need to be considered. Genetic and environmental factors are likely to play a role in determining susceptibility to autoimmunity.

Genetic elements, particularly those governed by HLA genes can influence the susceptibility of individuals to disease in many ways. First, different HLA haplotypes may affect the composition of the T cell repertoire, the relative expression of Vβ families, and the deletion or leakiness of autoreactive T cell clones. Second, certain HLA haplotypes may be more efficient than others in presenting autoimmune epitopes. Sercarz and his colleagues [54, 55] have suggested that infection with certain organisms and presentation of a dominant microbial protein epitope that is crossreactive with a cryptic self epitope can trigger autoreactive T cells. Of interest in this regard are our results showing that pep M5-primed, but not anti-CD3-primed, T cells proliferate in response to myosin. These data indicate that stimulation with pep M5 enriches for myosin responsive T cells. Further studies are required to characterize human T cell epitopes shared between myosin and M protein. We have recently localized a region within the pep M5 molecule that is important for its mitogenic activity. This region harbors a sequence motif, QKSKQXXGAXKQEL, that is shared by other superantigens [56], and the QKSKQ epitope which is the major epitope recognized by crossreactive anti-myosin antibodies found in sera of rheumatic heart disease patients [57]. Thus the superantigenicity of the M protein, coupled to the presence of molecular mimicry epitopes may potentiate immune responses directed to cryptic autoimmune epitopes. This new data raise exciting possibilities regarding the role of the M protein in the pathogenesis of rheumatic heart disease because it provides a link between infection, superantigenicity, molecular mimicry, and the response to self proteins.

Cytokine responses to infectious agents are likely to play an important role in affecting the susceptibility of certain individuals to disease. Abnormalities in cytokine production have been reported for a number of autoimmune diseases [24, 58] including rheumatic heart disease [59], but it is not clear whether these abnormalities are a cause or consequence of these disorders. Elevated levels of certain cytokines such as interferon-γ (INF-γ) can cause upregulation of HLA molecules and the induction of HLA class II expression by cells that normally do not express these molecules. This can lead to presentation of sequestered self antigens resulting in activation of autoreactive T cells. The quality and/or quantity of cytokines triggered by microbial proteins varies from one individual to another and can be influenced by concomitant infection with another co-pathogen. Recent studies from our laboratory have shown that pep M5 is a potent stimulator of cytokine production, but that this activity requires the participation of T cells [43]. The requirement for T cells could be substituted by addition of exogenous IFN-γ. Pep M5 induced the release of IFN-γ, TNF-α, IL-1 and IL-6 by PBMC [43]. Addition of excess IFN-γ to cultures of pep M5 and PBMC resulted in sustained higher levels of cytokines at 72 hours. Interestingly, TNF-β production only occurred in the presence of pep M5 and exogenous IFN-γ. The ability of IFN-γ to augment the production of cytokines, nitric oxide, and increase expression of cell adhesion molecules is bound to exacerbate harmful immune responses. Exposure, at the time of bacterial infection, to co-pathogens such as viruses which are known to induce secretion of high levels of IFN-γ could conceivably alter the quality, magnitude and kinetics of release of inflammatory cytokines thereby increasing the risk of developing autoimmunity.

In summary, the discovery of the mechanism by which superantigens arouse the immune system has provided strong support for the idea that infection with certain pathogens can trigger autoimmunity. However, when considering factors that determine susceptibility to autoimmunity, no single factor should be considered alone; instead, the context in which the host encounters the pathogen is a crucial predisposing factor and this may be influenced by host genetic factors and by environmental factors such as a concomitant exposure to a co-pathogen. The advancement in scientific methodologies, in particular the development of PCR technology, allows us for the first time to directly investigate the role of infection in triggering these diseases. Continued investigation of clinical material from affected individuals will help to unravel the infectious etiologies of autoimmunity and will provide the necessary insights to begin effective intervention in these devastating disorders.

References

1. Vaughan JH. Infection and rheumatic diseases: a review. Bull Rheum Dis 1990;39:1-8.
2. Schwartz BD. Infectious agents, immunity, and rheumatic diseases. Arthritis Rheum 1990;33:457-65.
3. Smiley JD, Hoffman WL. The role of infections in the rheumatic diseases: molecular mimicry between bacterial and human stress proteins? Am J Med Sci 1991;301:138-49.
4. Cohen IR, Young DB. Autoimmunity, microbial immunity, and the immunological homonuculus. Immunol Today 1991;12:154-158.
5. Rose NR, Neumann DA, Herskowitz A. Coxsackie virus myocarditis. Adv Intern Med 1992;37:411-29.
6. Moreland LW, Koopman WJ. Infection as a cause of reactive arthritis, ankylosing spondylitis, and rheumatic fever. Curr Opin Rheumatol 1992;4: 534-42.
7. Kotb M. Role of superantigens in the pathogenesis of infectious diseases and their sequelea. Curr Opinion Infect Dis 1992;5:364-74.
8. Nygard NR, Schwartz BD. Infection and autoimmunity. Adv Intern Med 1993;38:337-59.
9. Lotz M, Roudier J. Epstein-Barr virus and rheumatoid arthritis: cellular and molecular aspects. Rheumatol Int 1989;9:147-52.
10. Sulitzeanu D, Anafi M. EBV, molecular mimicry and rheumatoid arthritis: a hypothesis. Immunol Lett 1989;20:89-91.
11. Schattner A, Rager ZB. Virus-induced autoimmunity. Rev Infect Dis 1990;12:204-22.
12. Foulis AK. Does viral infection initiate autoimmunity in type I diabetes? J Autoimmun 1990;3:21-6.
13. Wolfgram LJ, Rose NR. Coxsackie virus infection as a trigger of cardiac autoimmunity. Immunol Res 1989;8:61-80.
14. Morrow WJ, Isenberg DA, Sobol RE, Stricker RB, Kieber ET. AIDS virus infection and autoimmunity: a perspective of the clinical, immunological, and molecular origins of the autoallergic pathologies associated with HIV disease. Clin Immunol Immunopathol 1991;58:163-80.
15. Barnett LA, Fujinami RS. Molecular mimicry: a mechanism for autoimmune injury. Faseb J 1992;6:840-4.
16. Friedman SM, Posnett DN, Tumang JR, Cole BC, Crow MK. A potential role for microbial superantigens in the pathogenesis of systemic autoimmune disease. Arthritis Rheum 1991;34:468-80.
17. Inman RD. Infectious etiology of rheumatoid arthritis. Rheum Dis Clin North Am 1991;17:859-70.
18. Lydyard PM, Rook GA, Tsoulfa G, Sharif M, Smith M. Is there a role for mycobacteria in the etiopathogenesis of rheumatoid arthritis? Immunol Rev

1991;121:137-54.

19. Trull AK, Ebringer R, Panayi GS, Colthorpe D, James DC, Ebringer A. IgA antibodies to *Klebsiella pneumoniae* in ankylosing spondylitis. Scand J Rheumatol 1983; 12: 249-53.
20. Davis P. Viral infections, acquired immunodeficiency syndrome, and rheumatic diseases. Curr Opin Rheumatol 1992; 4: 529-33.
21. Lane HC, Fauci AS. Immunologic abnormalities in the acquired immunodeficiency syndrome. Annu Rev Immunol 1985; 3: 477-500.
22. Steere AC. Current understanding of Lyme disease. Hosp Pract (Off Ed) 1993; 28: 37-44.
23. Granfors K. Do bacterial antigens cause reactive arthritis? Rheum Dis Clin North Am 1992; 18: 37-48.
24. Feldmann M. Molecular mechanisms involved in human autoimmune diseases: relevance of chronic antigen presentation. Class II expression and cytokine production. Immunol Suppl 1989; 2: 66-71.
25. Brennan FM, Feldmann M. Cytokines in autoimmunity. Curr Opin Immunol 1992; 4: 754-9.
26. Dinarello CA. Inflammatory cytokines: interleukin-1 and tumor necrosis factor as effector molecules in autoimmune diseases. Curr Opin Immunol 1991; 3: 941-8.
27. White J, Herman A, Pullen AM, Kubo R, Kappler JW, Marrack P. The V beta-specific superantigen staphylococcal enterotoxin B: stimulation of mature T cells and clonal deletion in neonatal mice. Cell 1989; 56: 27-35.
28. Herman A, Kappler JW, Marrack P, Pullen AM. Superantigens: Mechanism of T-cell stimulation and role in immune responses. Annu Rev Immunol 1991; 9: 745-72.
29. Paliard X, et al. Evidence for the effects of a superantigen in rheumatoid arthritis. Science 1991; 253: 325-9.
30. Howell MD, et al. Limited T-cell receptor β-chain heterogeneity among interleukin 2 receptor-positive synovial T cells suggests a role for superantigen in rheumatoid arthritis. Proc Natl Acad Sci USA 1991; 88: 10921-25.
31. Sottini A, Imberti L, Gorla R, Cattaneo R, Primi D. Restricted expresssion of T cell receptor Vβ but not V genes in rheumatoid arthritis. Eur. J. Immunol 1991; 21: 461-6.
32. Kotzin BL, et al. Preferential T-cell receptor beta-chain variable gene use in myelin basic protein-reactive T-cell clones from patients with multiple sclerosis. Proc Natl Acad Sci USA 1991; 88: 9161-5.
33. Sumida T, et al. T cell receptor repertoire of infiltrating T cells in lips of Sjogren's syndrome patients. J Clin Invest 1992; 89: 681-5.
34. Marguerie C, Lunardi C, So A. PCR-based analysis of the TCR repertoire in human autoimmune diseases. Immunol Today 1992; 13: 336-8.
35. Stollerman GH. Rheumatogenic group A streptococci and the return of

rheumatic fever. Adv Intern Med 1990; 35: 1-25.
36. Bisno A. Group A Streptococcal infections and acute rheumatic fever. New Engl J Med 1991; 325: 783-93.
37. Zabriskie JB. Rheumatic fever: a model for the pathological consequences of microbial-host mimicry. Clin Exp Rheumatol 1986; 4: 65-73.
38. Beachey EH, Bronze M, Dale JB, Kraus W, Poirier T, Sargent S. Protective and autoimmune epitopes of streptococcal M proteins. Vaccine 1988; 6: 192-6.
39. Cunningham MW, et al. Human monoclonal antibodies reactive with antigens of the group A Streptococcus and human heart. J Immunol 1988; 141: 2760-6.
40. Tomai M, Kotb M, Majumdar G, Beachey EH. Superantigenicity of streptococcal M protein. J Exp Med 1990; 172: 359-62.
41. Tomai MA, Aelion JA, Dockter ME, Majumdar G, Spinella DG, Kotb M. T cell receptor V gene usage by human T cells stimulated with the superantigen streptococcal M protein. J Exp Med 1991; 174: 285-8.
42. Tomai M, Schlievert PM, Kotb M. Distinct T Cell receptor Vβ Gene usage by human T lymphocytes stimulated with the streptococcal pyrogenic exotoxins and M protein. Infect Immun 1992; 60: 701-5.
43. Kotb M, et al. Temporal relationship of cytokine release by peripheral blood mononuclear cells stimulated by the streptococcal superantigen pep M5. Infect Immun 1993; 61: 1194-1201.
44. Le Gros HL, Aelion J, Geller A, Watanabe-Ohnishi R, Kotb M. In final preparation.
45. Shiverick KT, Thomas LL, Alpert NR. Purification of cardiac myosin application to hypertrophied myocardium. Biochemica et Biophysica Acta 1975; 393: 124.
46. Choi Y, Kotzin B, Herron L, Callahan J, Marrack P, Kappler J. Interaction of Staphylococcus aureus toxin "superantigens" with human T cells. Proc Natl Acad Sci USA. 1989; 86: 8941-5.
47. Kotb M, Courtney H, Dale JB, Beachey EH. Cellular and biochemical responses of human T lymphocytes stimulated by streptococcal M proteins. J Immunol 1989; 142: 966-70.
48. Beachey EH, Majumdar G, Tomai M, Kotb M. Molecular aspects of autoimmune responses to streptococcal M proteins. In: Advances in Host Defense Mechanisms (Gallin and Fauci, eds.) Raven Press. pp 83-96. 1990; 2: 83-95.
49. Watanabe-Ohnishi R, Tomai M, Aelion J, Kotb M. Characterization of unique human TCR-Vβ specificities for a family of streptococcal superantigens represented by rheumatogenic serotypes of M protein. 1993; (submitted).
50. Fischetti VA, Jones KF, Hollingshead SK, Scott JR. Structure, function, and genetics of streptococcal M protein. Rev Infect Dis 1988; 10: 356-9.

51. Fischetti M. Streptococcal M protein. Scientific Amer 1991; June 1991: 58-65.
52. Kisher K, Cuningham MW. Myosin: a link between streptococci and heart. Science 1985; 227: 413-5.
53. Stollerman GH. Streptococci and rheumatic heart disease. In: " The role of microorganisms in non-infectious diseases. (R.R.P. de Vries, I.R. Cohen, and J.J. van Rood, eds.) Springer-Verlag, New York. 1990; pp. 9-23.
54. Gammon G, Sercarz EE, Benichou G. The dominant nonself and the cryptic self: shaping the autoreactive T-cell repertoire. Immunol Today 1991; 12: 193-5.
55. Sercarz EE, Lehmann PV, Ametani A, Benichou G, Miller A, Moudgil K. Dominance and crypticity of T cell antigenic determinants. Annu Rev Immunol 1993; 11: 729-66.
56. Wang B, Schlievert PM, Gaber AO, Kotb M. Localization of an immunoloigically functional region of the streptococcal superantigen, pep M5. J Immunol 1993; (in press): .
57. Cunningham MW, McCormack JM, Fenderson PG, Ho MK, Beachey EH, Dale JB. Human and murine antibodies cross-reactive with streptococcal M protein and myosin recognize the sequence GLN-LYS-SER-LYS-GLN in M protein. J Immunol 1989; 143: 2677-83.
58. Eastgate JA, Symons JA, Wood NC, Grinlinton FM, Duff GW. Correlation of plasma interleukin-1 levels with disease activity in rheumatoid arthritis. Lancet 1988; 2: 706-8
59. Miller LC, et al. Cytokines and immunoglobulin in rheumatic heart disease: production by blood and tonsillar mononuclear cells. J Rheumatol 1989; 16: 1436-42.

INFECTION AND AUTOIMMUNITY: POSSIBLE CONTRIBUTION OF HEAT SHOCK PROTEINS

Stefan H.E. Kaufmann
Department of Immunology
University of Ulm
Albert-Einstein-Allee 11
D-7900 Ulm
GERMANY

Introduction

The superfamily of heat shock proteins (hsp) comprises a group of polypeptides which are synthesized at increased levels under various assaults [1]. Such insults not only include heat, the most widely used stress stimulus used in the laboratory, but also exposure to heavy metals, oxidants, infectious agents, and other assaults cells have to face while living. The hsp are remarkable in that they are ubiquitous, having been identified in every cell. Furthermore, hsp are highly conserved with respect to their structure, function and regulation. Because of their uniform appearance, hsp pose particular problems to the host immune system. Using hsp as target antigens would allow the immune system to identify stressed cells independent from the stress stimulus, such as infection, inflammation, transformation, trauma, or senescence. On the other hand, the high conservation raises a serious problem because it could deviate the reaction against infectious invaders into a response against the host. Although such a response needs not necessarily lead to autoimmune disease, the potential exists that it may proceed in such a detrimental direction. It is the purpose of this review to summarize recent findings related to the potential impact of hsp on autoimmune disease with an emphasis on the T-cell response.

Hsp: Function and Regulation of Gene Expression

Hsp have been grouped into distinct families according to their size. Members of a given family often perform similar functions and share a high degree of sequence homology [1]. Thus, members of the hsp 60 and the hsp 70 families possess a high affinity to hydrophobic polypeptide regions and hence are capable of interacting with unfolded proteins. This allows hsp 60 and hsp 70 cognates to promote the folding of newly synthesized proteins, the unfolding of proteins prior to their

A. G. Dalgleish et al. (eds.), The Impact of Biotechnology on Autoimmunity, 51–59.

translocation through membrane barriers and the assembly and disassembly of multimeric protein complexes [2]. In principle, this is true for both the prokaryotic and the eukaryotic members. The similarity, however, is not restricted to function, but also extends to structure. Bacterial and mammalian cognates of the hsp 60 family, for example, share more than 50% sequence homology with each other [1]. Hsp functions performed by various hsp families are depicted in Figure 1 [1].

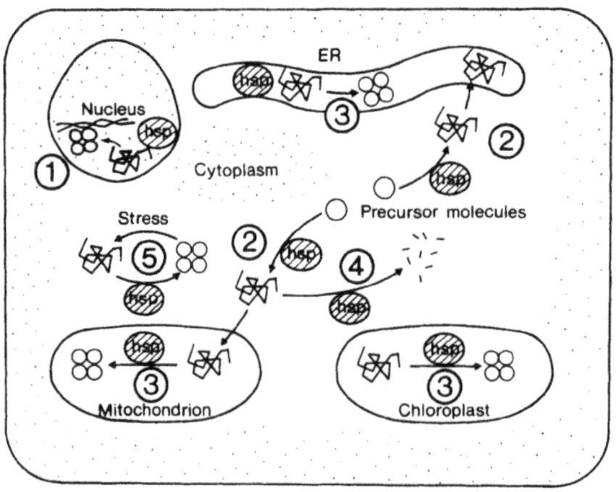

Figure 1. Hsp functions in normal and stressed cells. Some hsp are involved in the regulation of steroid receptors with nuclear DNA (1). Others promote folding of newly synthesized proteins and unfolding of proteins prior to translocation through membranes (2). Subsequent to membrane passage, proteins are again folded and assembled into oligomeric complexes with the help of hsp (3). Other hsp seem to contribute to protein degradation in lysosomal compartments (4). Under stress, oligomeric protein complexes disassemble and proteins unfold; this process, at least partially, is prevented or reversed by hsp (5). From [1].

The rapid increase of hsp synthesis following exogenous stress stimuli indicates a stringent regulation of hsp gene expression. Because of the great variety of assaults causing hsp synthesis, including physical (e.g., heat) and chemical stimuli (e.g., reactive oxygen metabolites, ROM), the identity of the transmitter system which senses and transmits exogenous signals into the cell remained a mystery for a long time. More recently, evidence has been presented that hsp 70 itself fulfills such a function [3,4]. Again, this is not only true in eukaryotes (see Figure 2), but also in prokaryotes.

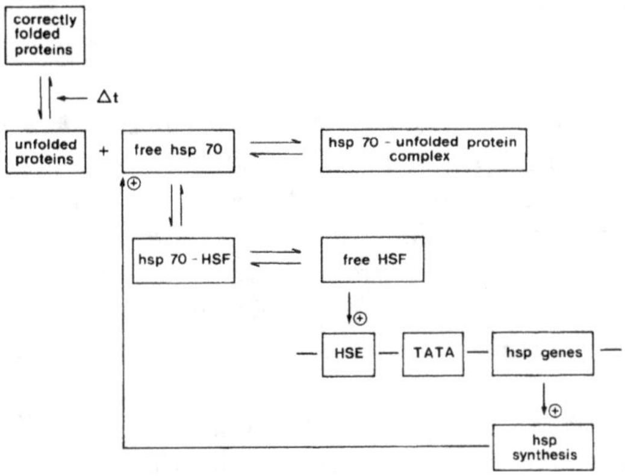

Figure 2. Autoregulation of the stress response by hsp 70. In cells, an equilibrium between unfolded and correctly folded proteins exists. Under various stress situations (e.g., heat, DT) the proportion of unfolded proteins increases. These unfolded proteins interact with free hsp 70 to form complexes. Hsp 70 not only interacts with unfolded proteins, but also with its heat shock factor (HSF). Reducing the amount of free hsp 70 reduces the proportion of hsp 70 HSF complexes, thus increasing the levels of free HSF. Free HSF forms a trimer that can bind to a specific sequence in the heat shock gene promoter (the heat shock element, HSE). The interaction between HSF and HSE results in transcriptional activation of heat shock genes, thus causing increased hsp synthesis. After [3] and [4].

Hsp and Autoimmune Disease

Two major lines of evidence suggest that hsp represent preferred target antigens of various forms of autoimmune disease. First, increased hsp synthesis has been observed in cells within autoimmune lesions. Second, T cells and antibodies with specificity for autologous hsp have been identified in numerous autoimmune diseases. In the following, autoimmune diseases in which hsp-reactive T cells had been identified, will be briefly summarized.

Adjuvant arthritis: In certain strains of rats a chronic inflammatory disease resembling rheumatoid arthritis can be induced by the injection of complete Freund's adjuvant which contains killed mycobacteria. T cells reactive with mycobacterial hsp 60 are capable of transferring disease [5]. They are specific for an amino acid (aa) sequence which is not particularly homologous with its mammalian counterpart. It is, however, possible that the T cells fortuitously cross-react with proteoglycans. Importantly, administration of mycobacterial hsp 60 prevents subsequent induction of adjuvant arthritis [5].

Insulin-dependent diabetes mellitus (IDDM) in non-obese diabetic (NOD) mice: In certain mouse strains an IDDM-like disease develops spontaneously. T cells from such mice transfer disease and these T cells recognize an aa stretch shared by the mycobacterial and the mammalian hsp 60 [6]. In this case, therefore, cross-reactivity between microbial and host hsp may provide a link between protection and autoimmune disease. Again, in this model, prior administration of hsp 60 or the relevant peptide prevents development of autoimmune disease.

Reactive arthritis: T cells with specificity for mycobacterial hsp 60 have been isolated from synovial lesions [7,8]. Some of these T cells are specific for a non-conserved microbial epitope, whereas others are specific for a conserved epitope shared by the mycobacterial and human hsp 60 cognates. However, more recent findings indicate that the frequency of such T cells is rather low [7].

Rheumatoid arthritis: Although T cells with specificity for hsp 60 have been described, more recent studies argue against a major role of hsp 60 in rheumatoid arthritis of adults [9,10]. T cells with specificity for a shared epitope of mycobacterial and human hsp 60, however, may play some role in the juvenile arthritis form [10].

IDDM: The relevance of hsp 60-specific T cells to IDDM remains questionable despite some arguments implicating this protein in disease development [11]. Glutamic acid decarboxylase appears to be a major autoantigen in IDDM [12]. Because some homology between human hsp 60 and glutamic acid decarboxylase has been observed, mimicry between these two proteins may contribute to the disease [13].

Multiple sclerosis (MS): γ/δ T cells have been identified in chronically demyelineated lesions and in acute MS plaques [14,15]. Furthermore, elevated hsp levels have been found in these lesions suggesting some role for hsp 60-reactive γ/δ T cells in disease development.

Mechanisms

In summary, at present the issue of hsp as major target antigen of autoimmune disease remains controversial. In particular, cross-reactivity between bacterial and mammalian hsp cognates seems to provide an insufficient, though partially valid, explanation. The following alternatives which are not mutually exclusive may be of relevance to this issue:

(1) Cross-reactivity between microbial and host hsp cognates.
(2) Mimicry between microbial hsp and unrelated host molecules.
(3) Self hsp as targets of T cells with regulatory or surveillance functions.
(4) Mimicry between host hsp and unrelated host molecules.

Without doubt, hsp represent dominant target antigens of many bacterial infections [1,10]. The stress encountered by bacteria inside host cells, particularly professional phagocytes, causes increased hsp synthesis in the microbial predator which may promote its intracellular survival [16]. The mere abundance of microbial hsp may then be responsible for the high frequency of T cells with specificity for this particular type of proteins. Conversely, infection causes stress in the affected host cell, thus also leading to increased hsp synthesis. Using peptides representing conserved regions of hsp 60, autoreactive T cells with specificity for shared epitopes of the bacterial and human cognates have been identified [17]. In a mouse model, CD8 T cells with specificity for mycobacterial hsp 60 have been shown to recognize stressed host cells [18]. Upon stimulation with various stress stimuli, hsp 60 synthesis is increased in these targets [19,20]. Addition of specific anti-sense oligodeoxynucleotides interferes with neosynthesis of hsp 60 and also inhibits lysis of stressed cells by the hsp 60-specific CD8 T cells [20]. These findings indicate that stressed cells, indeed, present epitopes derived from endogenous hsp 60 during stress. Furthermore, they suggest the feasibility of inhibiting autoimmune destruction with anti-sense oligodeoxynucleotides. These findings, of course, do not exclude the possibility that T cells with specificity for microbial hsp 60 recognize unrelated host proteins.

Several lines of evidence suggest that γ/δ T lymphocytes have a particular predilection for hsp. Thus, a high frequency of T cells from primary or secondary lymphoid organs recognize a region of hsp 60 which is partially shared by the mycobacterial and murine cognates [21,22]. Furthermore, murine γ/δ T cells recognize stressed host cells [23]. In the human system, γ/δ T cells with specificity for hsp 60 have been isolated from healthy individuals as well as from patients [24, 25]. Infection of mice with intracellular bacteria such as *Listeria monocytogenes* results in the rapid influx of γ/δ T cells at the site of microbial implantation [26]. Besides performing a transient effector function in antibacterial defence, these γ/δ T cells seem to perform regulatory functions, in addition. Mice in which the δ-chain of the T-cell receptor had been deleted by homologous recombination and which, therefore, are devoid of any γ/δ T cells, were infected with *L.monocytogenes* [27]. In these mice, huge abscesses developed, whereas in *L.monocytogenes*-infected controls granulomas were formed. These findings suggest control of α/β T cells by γ/δ T cells in situ. In another set of experiments, normal mice were treated with the monoclonal antibody (mAb) GL3 to down-modulate γ/δ T-cells functions [28]. α/β T cells from these mice proliferated *in vitro* and produced significant amounts of IL-2 in the absence of foreign antigens. Furthermore, the α/β T cells from these mice developed into cytolytic T lymphocytes capable of lysing not only allogeneic but also syngeneic target cells. These findings suggest uncontrolled replication and differentiation of α/β T cells in the absence of γ/δ T cells. It, therefore, appears that γ/δ T cells control α/β T-cell activation and hence are an essential part of immune homeostasis. The link between hsp recognition by γ/δ T cells and these regulatory functions needs to be further analyzed. However, it is tempting to speculate that (i) hsp 60 serves as target antigen of regulatory γ/δ T cells and (ii) that disturbances in this regulatory cross-talk may facilitate development of autoimmune disease. A recent computer search has revealed that numerous regions of hsp 60 of >10 aa length share an appreciable similarity with unrelated host antigens which have been implicated in various autoimmune diseases [13]. It is, therefore, possible that T cells with specificity for self-hsp 60, once activated, cross-react with unrelated host molecules.

Concluding Remarks

The relation between hsp 60 and other hsp families to autoimmune disease remains an interesting, though unresolved, issue (Figure 3). Initially, it was thought that

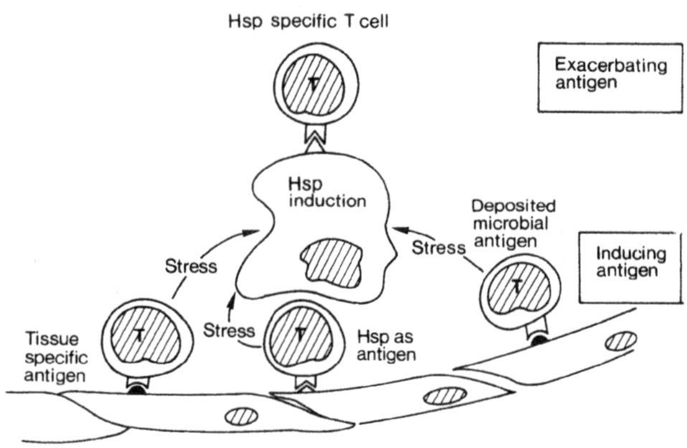

Figure 3. Three levels of interaction at which hsp may contribute to autoimmune disease.

(1) T cells which had been activated by hsp from infectious agents could induce an inflammatory reaction (i) because they interact with deposited microbial hsp; (ii) because they recognize regions shared by the microbial hsp with its autologous cognate or (iii) because they recognize tissue-specific antigens which, fortuitously, resemble microbial hsp. Alternatively, the inducing antigen is not a hsp, but a tissue-specific antigen. (2) In either case, such interactions cause stress in the lesion, thus increasing hsp synthesis and presentation of hsp-derived epitopes. In this case, hsp acts as exacerbating antigen. (3) Hsp may represent target antigens of regulatory T cells, particularly γ/δ T cells. Appearance of these hsp-reactive regulatory T cells at the site of inflammation would then reflect the failure of the immune system to downregulate the aggravated immune response.

cross-reactivity between microbial and host hsp cognates provides an essential link between infection and autoimmunity. An attractive aspect of this notion was the fact that the etiology was shifted from the microbial agent to a defined protein. In extension of this notion, T cells activated by microbial hsp may also cross-react with unrelated self proteins. In either case, hsp would function as inducing antigen. Alternatively, the T cells which induce autoimmune disease are specific for tissue antigen(s). The insults arising at the site of inflammation may then cause increased hsp synthesis attracting hsp-reactive T cells to the site of action. In this case hsp would act as exacerbating antigens. Finally, hsp may serve as target antigens of regulatory T cells. Then, identification of hsp-reactive T cells in autoimmune lesions would not mirror an effector function, but rather reflect the idle attempt to prevent an ongoing autoimmune process.

Acknowledgements

This work received financial support from the Sonderforschungsbereich 322. The expert secretarial assistance of Mrs. R. Mahmoudi is gratefully acknowledged.

References

1. Kaufmann SHE. Heat shock proteins and the immune response. Immunol Today 1990; 11:129-136.
2. Ellis RJ, van der Vies SM. Molecular chaperones. Ann Rev Biochem 1991; 60:321-347.
3. Morimoto RI. Cells in stress: Transcriptional activation of heat shock genes. Science 1993; 259:1409-1410.
4. Craig EA, Gross CA. Is hsp70 the cellular thermometer? Trends Biochem Sci 1991; 16:135-140.
5. van Eden W, Thole JE, Van der Zee R, et al. Cloning of the mycobacterial epitope recognized by T lymphocytes in adjuvant arthritis. Nature 1988; 331:171-173.
6. Elias D, Markovits D, Reshef T, Van der Zee R, Cohen IR. Induction and therapy of autoimmune diabetes in the non-obese diabetic (NOD/Lt) mouse by a 65-kDa heat shock protein. Proc Natl Acad Sci USA 1990; 87:1576-1580.
7. Life PF, Bassey EOE, Gaston HJS. T-cell recognition of bacterial heat shock proteins in inflammatory arthritis. Immunol Rev 1991; 121:113-135.
8. Hermann E, Lohse AW, Van der Zee R, et al. Synovial fluid-derived Yersinia-reactive T cells responding to human 65-kDa heat-shock protein and heat-stressed antigen-presenting cells. Eur J Immunol 1991; 21:2139-2143.
9. Res PCM, Schaar CG, Breedveld FC, et al. Synovial fluid T cell reactivity against the 65-kD heat-shock protein of mycobacteria in early chronic arthritis. Lancet 1988; II:478.

10. De Graeff-Meeder ER, Van der Zee R, Rijkers GT, et al. Recognition of human 60 kD heat shock protein by mononuclear cells from patients with juvenile chronic arthritis. Lancet 1991; 337:1368-1372.
11. Jones DB, Hunter NR, Duff GW. Heat-shock protein 65 as a b-cell antigen of insulin-dependent diabetes. Lancet 1990; 336:583-585.
12. Baekeshow S, Jan-Aanstoot H, Christgau S, et al. Identification of the 65 K autoantigen in insulin-dependent diabetes as the GABA-synthesizing enzyme glutamic acid decarboxylase. Nature 1990; 347:151-156.
13. Jones DB, Coulson AFW, Duff GW. Sequence homologies between hsp60 and autoantigens. Immunol Today 1993; 14:115-118.
14. Wucherpfennig KW, Newcombe J, Li H, Keddy C, Cuzner ML, Hafler DA. γ/δ T-cell receptor repertoire in acute multiple sclerosis lesions. Proc Natl Acad Sci USA 1992; 89:4588-4592.
15. Selmaj K, Brosnan CF, Raine CS. Colocalization of lymphocytes bearing γ/δ T-cell receptor and heat shock protein hsp65^{+} oligodendrocytes in multiple sclerosis. Proc Natl Acad Sci USA 1991; 88:6452-6456.
16. Kaufmann SHE. Heat shock proteins and pathogenesis of bacterial infections. Springer Semin Immunopathol 1991; 13:25-36.
17. Munk ME, Schoel B, Modrow S, Karr RW, Young RA, Kaufmann SHE. Cytolytic T lymphocytes from healthy individuals with specificity to self epitopes shared by the mycobacterial and human 65 kDa heat shock protein. J Immunol 1989; 143:2844-2849.
18. Koga T, Wand-Württenberger A, DeBruyn J, Munk ME, Schoel B, Kaufmann SHE. T cells against a bacterial heat shock protein recognize stressed macrophages. Science 1989; 245:1112-1115.
19. Kaufmann SHE, Schoel B, Koga T, Wand-Württenberger A, Munk ME, Steinhoff U. Heat shock protein 60: Implications for pathogenesis of and protection against bacterial infections. Immunol Rev 1991; 121:67-90.
20. Steinhoff U, Zügel U, Hengel H, Rösch R, Munk ME, Kaufmann SHE. Anti-sense oligodeoxynucleotides to heat shock protein 60 interfere with autoimmune destruction of stressed host cells by cytolytic T lymphocytes. Submitted
21. Born W, Hall L, Dallas A, Boymel J, Shinnick T, Young D, Brennan P, O'Brien R. Recognition of a peptide antigen by heat shock reactive γ/δ T lymphocytes. Science 1990; 249:67-69.
22. Fu Y-X, Cranfill R, Vollmer M, Van der Zee R, O'Brien RL, Born W. In vivo response of murine γ/δ T cells to heat shock protein-derived peptide. Proc Natl Acad Sci USA 1993; 90:322-326.
23. Rajasekar R, Sim GK, Augustin A. Self heat shock and γ/δ T-cell reactivity. Proc Natl Acad Sci USA 1990; 87:1767-1771.
24. Holoshitz J, Koning F, Coligan JE, DeBruyn J, Strober S. Isolation of CD4^{-} CD8^{-} mycobacteria-reactive T lymphocyte clones from rheumatoid arthritis synovial fluid. Nature 1989; 339:226-229.

25. Haregewoin A, Soman G, Hom RC, Finberg RW. Human γ/δ^{+} T cells respond to mycobacterial heat-shock protein. Nature 1989; 340:309-312.

26. Hiromatsu K, Yoshikai Y, Matsuzaki G, et al. A protective role of γ/δ T cells in primary infection with Listeria monocytogenes in mice. J Exp Med 1992; 175:49-56.

27. Mombaerts P, Arnoldi J, Russ F, Tonegawa S, Kaufmann SHE. Differential roles of α/β and γ/δ T cells in immunity against an intracellular bacterial pathogen. Nature, in press

28. Kaufmann SHE, Blum C, Yamamoto S. Crosstalk between α/β T cells and γ/δ T cells in vivo: Activation of α/β T cell responses after γ/δ T cell modulation with the monoclonal antibody GL3. Proc Natl Acad Sci USA 1993; in press.

CHRONIC INFECTION AND AUTOIMMUNITY

David A. Isenberg
Bloomsbury Rheumatology Unit
Department of Medicine
University College London Medical School
London
UNITED KINGDOM

Introduction

There seems little doubt that the immune system has developed as an essential protective mechanism to ensure the survival of any species. In the human species it is particularly adept at providing protection against infectious organisms, whether viral or bacterial. However the complexity of the immune system is such that malfunctions are almost inevitable during the lifetime of any individual. A combination of these malfunctions and other factors, such as the hormonal environment and exposure to particular environmental agents, leads to the development of autoimmunity. This may manifest as raised levels of autoantibodies, such as rheumatoid factor or anti-single stranded DNA antibodies unaccompanied by clinical features. However as many as 1 in 40 individuals develop a clinically overt autoimmune disease ranging from those which are organ specific such as thyrotoxicosis, to multi-system organ diseases such as systemic lupus erythematosus (SLE). There are thus intimate links between infectious organisms and autoimmunity which may be summarized under the following headings (the question of autoimmune disease-initiation/exacerbation [no. 3 below] is not discussed in this article):

1. Autoantibodies appearing after infectious diseases
2. Antibacterial antibodies associated with autoimmune disease
3. Autoimmune disease-initiation/exacerbationfollowing infection
4. Molecular mimicry
5. Clues from studies of autoantibody specificities and origins
6. Idiotypic considerations
7. Heat shock proteins
8. Glycosylation abnormalities of IgG

A. G. Dalgleish et al. (eds.), The Impact of Biotechnology on Autoimmunity, 61–66.

AUTOANTIBODIES APPEARING AFTER INFECTIOUS DISEASES

Many viral bacterial parasitic infections are associated with the appearance of autoantibodies. Thus Dziarsky [1] first showed that infecting healthy animals with viral or bacterial products led to the development of rheumatoid factor, anti-DNA, and anti-red cell antibodies. Furthermore, incubation of the normal peripheral blood lymphocytes with lipopolysaccharides also induces the production of autoantibodies. However a profile of autoantibody production in infectious diseases reveals that certain types such as rheumatoid factors, antisingle stranded DNA and antipoly (ATP-ribose) antibodies are often found, but antibodies to the so-called extractable nuclear antigens (anti-Sm, anti-RNP, anti-Ro, anti-La) are extremely uncommon in infectious diseases, though relatively frequently found in SLE [2].

ANTIBACTERIAL ANTIBODIES ASSOCIATED WITH AUTOIMMUNE DISEASES

The possible mechanisms for autoantibody production in infectious diseases include (i) polyclonal activation of B lymphocytes by constituents of the invading organism, and (ii) following an invasion by the organism damage to self occurs and this leads to changes in the target self antigens, (iii) specific induction by the invading organism of antilymphocytic antibodies which in turn suppress lymphocyte subsets such as T suppressors.

Raised antibody levels to various streptococcal cell wall antigens have been found in patients with adult onset rheumatoid arthritis and juvenile chronic arthritis. Intriguingly, an increased level of antibodies to *Proteus mirabilis* have been found in patients with rheumatoid arthritis, which has led Ebringer et al. [3] to propose that this may be an important trigger in patients with this condition.

MOLECULAR MIMICRY

This term, first coined by George Snell, was initially used to suggest that the MHC and viruses (especially retroviruses) might encode similar antigens which allow the host to regard an infecting virus as self and thus forego an immune response. There are now numerous examples of molecular mimicry in human biology. Examples include: proteins present in neurones which are affected during Chagas disease and the proteins of *Trypanosoma cruzi*; components of cardiac tissue and streptococcal polysaccharides; and CNS antigens and epitopes on *Borrelia burgdorferi* (the organism associated with Lyme arthritis). However as Horsfall [4] has proposed, in order to determine truly relevant molecular mimicry of B cell epitopes a number of criteria need to be satisfied including: a) sequence similarity, (applicable to linear epitopes); b) antisera should bind autoantigen and the crossreactive sequence; c) affinity purified antibody should bind both proteins; d) each antigen or peptide sequence should inhibit the binding autoantibody to the other; e) immunization with synthetic peptide corresponding to cross reactive epitopes should elicit autoantibodies.

CLUES FROM STUDIES OF AUTOANTIBODY SPECIFICITIES AND ORIGINS

It was once thought that the immune system could not permit the production of autoantibodies in normal circumstances, as they might be potentially pathogenic. It is now recognized however that so-called natural antibodies do exist [5]. They are often polyreactive low-affinity IgM immunoglobulins, able to react with both self and non-self antigens. They almost certainly play a regulatory role in the immune system, they are idiotypically interconnected and it has been suggested that a likely role for these autoantibodies is as a first line of defense against microbial agents. It is also thought likely that the source of these antibodies are CD5+ B cells.

Links between natural autoantibodies and more pathogenic antibodies are also being studied by analysis of the amino acid sequences which constitute the antibodies and the germline genes which encode them. Studies of both rheumatoid factors and anti-DNA antibodies have suggested that germline genes may encode autoantibodies. Thus the germline gene sequence V_H26 (a member of the V_H3 family) has been shown to encode a series of antibodies which both carry the 16/6 idiotype (see later) and also have DNA binding potential [6]. In the main however the sequence of the human DNA antibodies that have been determined do show mutations away from the germline gene sequences. A key question though is to determine whether there are particular areas in the amino acid sequence which determine the ability to bind DNA. An interesting example of this type of analysis which links a chronic infection (infectious mononucleosis) with a variety of autoantibodies and certain tumors has been the study of a particular germline gene segment designated V_H4-21. It is discussed in detail in the next section.

IDIOTYPIC CONSIDERATIONS

Although antibodies are characteristically defined by the antigens with which they bind, they may also be distinguished on the basis of epitopes present on either the heavy or light chains (or some combination of the two) in the hypervariable regions. These are designated idiotypes and are responsible for the induction of a further set of antibodies which recognize them as anti-idiotypes. The sharing of idiotypes by unrelated individuals carries the implication that the genes which encode the amino acids which constitute the idiotypes may also be shared. There are now several examples of idiotypes shared by both autoantibodies and antibacterial antibodies. Perhaps the best example is the 16/6 idiotype [7] which has been shown to be present on antibodies to DNA, anti-Ro, anti-Sm-RNP and anti-Klebsiella antibodies. In addition, the 3I idiotype like the 16/6 Id first identified on anti-DNA antibodies has been found on antipneumococcal idiotypic antibodies [8]. It seems inherently more likely that the genes which encode for these idiotypes have been preserved because they are concerned with protecting the species against infection than because they might form part of potentially pathogenic anti-DNA antibodies. However these idiotypes may in certain circumstances be a component in the pathogenesis of autoimmune diseases. 16/6 and 3I idiotypes have both for example been identified on immunoglobulins deposited in the renal lesions of lupus patients.

As referred to above, the V_H4-21 gene segment appears to be an important link between infection, autoimmunity and malignancy. This gene segment encodes for an idiotype known as the 9G4 Id, and was first identified on the naturally occurring monoclonal antibodies that bind the red blood cell antigen Ii. It is frequently present in the serum of patients with infectious mononucleosis (due to the Epstein-Barr virus). Subsequently this idiotope has been found on anti-DNA antibodies [9], rheumatoid factor, and most recently on a human monoclonal anti-myeloperoxidase antibody (Ehrenstein, Cambridge, Stevenson and Isenberg, unpublished observations). The idiotope is present in raised levels in up to 50% of SLE patients. In an analysis of the DNA antibodies from six lupus patients the idiotope was present on between 2% and 20% of these immunoglobulins.

HEAT SHOCK PROTEINS

This remarkable family of proteins is amongst the most highly conserved throughout evolution. They are subdivided into families according to molecular weight, have key physiological functions and help the survival of stressed cells. The physiological roles include the chaperoning of other proteins around the cytoplasm, the correct folding of proteins as they are formed from the messenger RNA code, and they also appear to be important in the normal functioning of a variety of enzymes. Links between heat shock protein expression and autoimmunity have been extensively explored in the last decade.

In particular efforts have been made to link the expression of hsp65 with rheumatoid arthritis. Although it has been shown that T cell clones which recognize a key epitope in hsp 65 (amino acids 180 to 188) can induce a form of adjuvant arthritis in predisposed strains, attempts to link hsp 65 with human rheumatoid arthritis have in general been unsuccessful [10]. More recently in our own laboratories, David Latchman and I have looked at the overexpression of hsp 90 with systemic lupus erythematosus. A summary of our findings follows: (a) hsp 90 levels are overexpressed in approximately 25% of SLE patients; (b) those patients most likely to overexpress hsp 90 have active disease especially involving the central nervous system and the heart and lungs; (c) patients who are A1 B8 DR3 are statistically significantly less likely to overexpress hsp 90; (d) an association between those patients who overexpress hsp 90 and those with the antiphospholipid antibody snydrome has been demonstrated; (e) in murine studies it has been shown that levels of hsp 90 in the splenic tissue of MRL/lpr mice (though not in other organs and tissues tested) are elevated prior to the onset of disease; (f) in humans the raised levels of hsp 90 or associated with increased transcription of hsp 90 β gene (but not the α gene). This subject is reviewed in detail elsewhere [11].

GLYCOSYLATION ABNORMALITIES OF IgG

The CH2 domains of IgG are bridged by bi-antennary oligosaccharide chains which normally terminate in galactose and sialic acid moieties. It has been shown that there is an age dependent variation in the number of chains which terminate prematurely, lacking all galactose units (known as Gal(0)) [12]. Allowing for age dependent phenomenon, a collaborative study of some 30 different diseases - between the department of Biochemistry at the University of Oxford and the Department of Rheumatology at the Middlesex Hospital - has shown that a restricted number of diseases are associated with raised levels of Gal(0). In particular, rheumatoid arthritis, tuberculosis and Crohn's disease [13]. The overwhelming majority of patients with these conditions when active have raised levels of Gal(0). Furthermore it has been shown that levels of Gal(0) fluctuate with disease activity, and levels of the lymphocyte galactosyl transferase which is responsible for inserting galactose into the oligosaccharide chain are also reduced in both B and T lymphocytes in patients with these conditions. The absence of galactose in the IgG molecules does have functional consequences for this immunoglobulin. It has thus been suggested that hypogalactosylation is an important component of the pathogenesis of these diseases.

Raised levels of Gal(0) have been shown to be clinically useful, when combined with the presence of rheumatoid factor in patients with early onset synovitis [14]. Thus it was shown that those individuals with raised Gal(0) levels and a positive rheumatoid factor had a 94% chance of going on to develop rheumatoid arthritis. In another study, however, it was shown that whereas the relatives of the patients with rheumatoid arthritis had normal levels of Gal(0) the spouses, i.e. those living with the patients, also tended to have raised levels. This would suggest that there is an important environment factor which is responsible for the increase.

Conclusion

These studies, derived from many different sources, all point to intimate connections between infection and immunity. As human beings we exist in an environment surrounded by potential pathogens. The sophistication of the immunological mechanisms that have evolved to protect us is remarkable but clearly not infallible. In certain circumstances, elements of our own immune system clearly conspire against us - the end result being the clinical expression of autoimmunity in a significant proportion of the human population.

References

1. Dziarsky R. Preferential induction of autoantibody secretion in polyclonal activation by peptidoglycan[s] and lipoplysaccharides I. In vitro studies. J Immunol 1982;128:1018-25.
2. Isenberg DA, Maddison P, Swana G, et al. Profile of autoantibodies in the serum of patients with tuberculosis, Klebsiella and other gram negative infections. Clin Exp Immunol 1987;67:507-15.
3. Ebringer A, Ptaszynska T, Corbett M, et al. Antibodies to Proteus in RA. Lancet 1985;2:305-7.
4. Horsfall AC. Molecular mimicry and antoantigens in connective tissue diseases. Mol Biol Reports 1991;16:139-48.
5. Shoenfeld Y and Isenberg DA, editors. Natural Autoantibodies Their Physiological Role and Regulatory Significance. Boca Raton: CRC Press: 1992.
6. Dersimonian H, Schwartz RS, Barrett KJ, Stollar BD. Relationship of human variable region heavy chain (VH) germ line genes to genes encoding anti-DNA autoantibodies. J Immunol 1987;139:2496-501.
7. Isenberg DA. Anti-DNA antibody idiotype 16/6 - a system of some importance. Ann De Med Interne 1990;141:57-63.
8. Grayzel A, Solomon A, Aranow C, Diamond B. Antibodies elicited by pneumococcal antigens bear an anti-DNA-associated idiotype. J Clin Invest 1991;87:842-6.
9. Isenberg DA, Spellerberg M, Williams W, Griffiths M, Stevenson F. Identification of a role for the 9G4 idiotype in systemic lupus erythematosus. Submitted.
10. Gaston JSH. Are hear shock proteins involved in autoimmunity? Int J Clin Lab Res 1992;22:90-4.
11. Isenberg DA. Heat shock proteins in autoimmune disease. In the heat of the fight? Eur J Nephrol, in press.
12. Parekh R, Isenberg DA, Roitt I, Dwek R, Rademacher T. Age related galactosylation of the N-linked oligosaccharides of human serum IgG. J Exp Med 1988;167:1731-6.
13. Parekh R, Isenberg DA, Rook G, Roitt I, Dwek R, Rademacher T. A comparative analysis of disease associated changes in the galactosylation of serum IgG. J Autoimm 1989;2:101-14.
14. Young A, Sumar N, Bodman K, Goyal S, Sinclair H, Roitt I, Isenberg DA. Agalactosyl IgG - an aid to differential diagnosis in early synovitis. Arthritis Rheum 1991;34:1425-9.

THE ROLE OF HLA MIMICRY BY HIV IN THE PATHOGENESIS OF AIDS

Angus Dalgleish, Louise Brown, Bernard Souberbielle, Michael Westby
St. George's Hospital Medical School
Division of Oncology
Cranmer Terrace
London SW17 0RE
UNITED KINGDOM

The pathogenesis of retrovirus induced disease is extraordinarily complex for such genetically small and similar organisms. For instance, oncogenic retroviruses can induce disease by 'taking' a cellular 'oncogene' from one cell host and placing it in another, by inserting their double stranded DNA copy next to a cellular oncogene and either activate it in cis, or inhibit normal regulatory suppressor signals. The first human retrovirus (HTLV-1) uses yet another mechanism inserting into the host DNA at random and turning on cellular genes in trans [1]. HIV is a lentivirus and as such a member of a group of viruses which, although they may be cytopathic *in vitro*, cause disease *in vivo* by stimulating an inappropriate and autodestructive immune response. (e.g. anaemia in the equine infectious anaemia virus, arthritis in the caprine arthritis encephalitis virus and neurological damage in the visna maeda virus). It is therefore possible that HIV and SIV may likewise induce their damage or disease by autoreactive mechanisms [2].

A review of the immune response to HIV infection essentially demonstrates that as the disease progresses, nothing remains normal! It is therefore hard to determine the cause and effect of the resultant immune suppression. Destruction of CD4 lymphocytes is not enough to account for many of the changes, the most important features of which are:

1) Panactivation of the immune system including CD4, CD8 and B cells. The activation is progressive with nearly all CD8 cells expressing the CD38 marker when CD4 cells are less than 50.
2) Antigen specific anergy which is selectively lost to self presented antigens and later to allopresented antigens.
3) Autoreactive CTL activity with loss of virus specific CTL activity as disease progresses (not seen in HIV infected chimpanzees who do not get ill).

As such the immune response is similar to conditions which non-specifically activate

A. G. Dalgleish et al. (eds.), The Impact of Biotechnology on Autoimmunity, 67–72.

the immune system such as superantigens (e.g. as in the murine AIDS model) or alloreactive disease also known as graft versus host disease. The fact that HIV gp120 binds CD4 and has more than one conserved region homologous to HLA led us to ask if HIV could functionally mimic HLA in that it could activate through the T-cell receptor in a similar manner to allogeneic HLA [3].

Evidence from Computer Modeling

Molecular modeling of the carboxy terminus suggested that gp120 was similar to HLA Class I (A_2) [4]. Although HLA sequence homologies are more similar to HLA Class II than Class I, HLA Class II was not crystallized and sequenced molecules were used for comparison.

The carboxy terminus of gp120 is homologous to HLA DR.B3. We therefore constructed peptides from this region and investigated whether these peptides would be recognized by alloactivated lymphocytes from a panel of uninfected donors. In collaboration with Mario Clerici and Gene Shearer of the NCI, NIH, we were able to show that alloreactive lymphocytes from most (about 80%) donors recognized 'self', cells plus the carboxy terminus peptide in the same manner as allogeneic target cells [5]. Using the peptide as an antigen we are looking for the induction of spontaneous proliferation using peripheral blood lymphocytes from donors with HLA types associated with rapid progression and comparing them with donors with HLA types associated with slow progression.

Evidence from Antibody Studies

While these studies were ongoing, Alberto Beretta and colleagues, from the San Raffaele Institute in Milan, characterized two monoclonal antibodies raised against HLA Class I molecules on activated cells which also recognized HIV gp120. Epitope mapping of these antibodies show that they recognize the same carboxy terminus region as the alloactivated used in our studies. In addition, they also recognize a YKYK motif also conserved in HIV and HLA sequences. Interestingly, their antibodies (which are against HLA Class I, type C) do not recognize the sequences between the YKYK and the carboxy terminus peptide used above, Beretta has suggested that the backbone of MHC mimicry acts as a decoy to prevent recognition of potentially protective epitopes [6]. Nevertheless, the antibody data is in keeping with the speculation that HIV could mimic HLA directly.

HLA Mimicry by Incorporation of Host MHC Molecules

In addition to the '3D' mimicry potential of gp120 (160), there also exists the probability that MHC molecules incorporated into the viral envelope from the host cell could stimulate through the T-cell receptor in an alloreactive manner. Several viruses have now been reported to incorporate MHC molecules into their envelope including viruses such as HTLV-1, which do not cause panactivation and immunosuppression in the same manner as HIV or SIV. However, the ability to bind CD4 in addition to incorporating host MHC may be crucially important. Recently Arthur et al. [7] have reported restricted incorporation of specific HLA molecules by a number of techniques. Early studies by one of us (AG Dalgleish, unpublished observation) suggested that the incorporated MHC molecules were shed after about 10-15 minutes. It is therefore not clear if this process would maintain allo MHC molecules long enough to drive the disease. However, HIV may alter self MHC to induce an inappropriate TCR signal in this model. Experiments from SIV infected macaques suggest that incorporated HLA molecules may be an important component of vaccines as uninfected human cells could protect against challenge with SIV genes in humans but not monkey cells. Some workers have found that the protection correlates with an anti-HLA response [8].

Peptides as Alloantigens

Analysis of HLA types associated with fast and slow progression from a number of sources show that whereas no one association with disease is consistently seen (this should be obvious when population HLA polymorphisms are considered) certain patterns do exist. For example, the HLA types associated with slow progression HLA type B52, B44, B27, B18, B21 and B57, DR BI (14a, 13a, 16 and 17.1) are all relatively rare HLA types. Fast HLA types such as B35, B8, B51, B7 and DR 2, 3, 5 and 11 are much more common and some are already known to be associated with autoimmune or autoreactive disease [2,3,9].

The sequences of HIV from the conserved regions are as follows:

```
247   Q C T H G I  K P I  V S T Q L L N G S L A E
                  (R)    (I)

432   Y A P P I

444   S N I  T G I L  T R D G G
                (L)(I)

468   G G G N M K D N W
               (D) (R)

479   E L Y K Y K V I
                     (V)

488   I E P L G I A P T K A K R R V V E R E K A
               (I)     (R)         (Q)
```

The region from which the carboxy peptide used in the alloreactive experiments described was derived viz.: T R A K R R V V E R E K R, is a mimic of the DR.B chain, which unlike the gp120 DR/DQ-like 247 sequence (which is non polymorphic in humans and chimpanzees) codes subtle differences between chimpanzees and man.

The common residues could allow self HLA Class I molecules to present HLA Class II sequences either as self sequences (in which case the immune system would be tolerant to activation) or as a foreign or allo HLA Class II peptide sequence. In theory, HLA Class I restricted presentation of an HLA Class II alloepitope could be mimicked by HIV in most humans but not in chimpanzees. Alloepitopes of HLA DR have been described between DR B3, 0101 and 0201 in that if the sequences between the conserved arginine anchors amino acid numbers (72 and 80) are compared then arginine substitution for glutamine can induce allocytotoxic reactions between HLA Class II subtypes. The HIV homologous sequences would allow similar opportunities for inducing such reactivity, the exact changes of which will depend on the HLA Class I molecule used (JA Habeshaw, personal communication).

Consequences of Alloreactivity

The biology of HIV infection *in vivo* is ideally suited for the induction of chronic activation of the immune system by either three dimensional HLA mimicry or by peptides. A major target apart from CD4 lymphocytes, are the antigen presenting cells of the lymph nodes and lymphoreticular organs [10]. These cells are not readily killed by HIV but instead are able to continually present the virus envelope as HLA mimic or 'allo' HIV peptides to trafficking lymphocytes. Only those individuals who possessed T lymphocytes which recognized these alloepitopes would respond. This response would be similar to that seen in chronic graft versus host disease (GVHD). Recently much has been made of the differentially T helper cells response in the outcome of disease. Progressive disease is associated with a Th2 response which is associated with an IL-4, IL-6 and TNFα production, whereas asymptomatic infection is associated with a Th1 predominant response (IL-2 and γIFN). It is of interest that GVHD is predominantly a Th2 cell response as is the murine 'AIDS' model which is induced by a defective murine leukemia virus [11,12]. The outcome on both 'MAIDS' and GVHD is crucially dependent upon the genetic background of the host [3].

Implications for Pathogenesis

The classical viewpoint of AIDS is that it is caused by a CD4 lymphocyte virus and progression is associated with hypervaribility evasion of the immune system and the establishment of a large virus load. Many immunologists suggest that the hypervariability of the virus leads to an activated and subsequently exhausted immune system. An alternative explanation is that HIV infects antigen presenting cells such as dendritic and monocyte macrophage populations thus rendering them defective with regards exogenous antigen presentation. Defective signalling could explain the activation of the CD45RA+ subset to the CD45RO+, or memory population, with a reduced return of RO+ cells to the RA+ pool. Such an imbalance would render the cells more susceptible to apoptosis by exogenous superantigens. It is likely that aspects of all these interpretations will turn out to play some role in the pathogenesis [13]. However, there is no obvious reason why these explanations should be affected by HLA restrictions. Here we have proposed that the activation and disease progression occurs in a non-random fashion as it is associated with known HLA types and, as such, HLA associated activation may not necessarily be as direct as proposed here [14].

The mechanism of pathogenesis proposed here will allow experiments to be designed to further define the basic principles. One obvious one is that thalidomide is very effective at suppressing a Th2 response (in contrast to cyclosporin A) and is currently the drug of choice in some bone marrow transplants units to GVHD [15]. We have therefore commenced a randomized study to see if thalidomide will prevent disease progression in infected people with elevated activation markers. Specific therapy could be aimed at tolerizing to the 'allo' epitope(s). Such approaches may then allow standard antivirus therapies and therapeutic vaccine strategies to be effective.

References

1. Dalgleish AG. Viruses and Cancer. In: Sikora K, Watson G, Evans G, eds. British Medical Bulletin 1991;1:21-46.
2. Dalgleish AG. The pathogenesis of AIDS: classical and alternative views. J Royal College of Physicians 1992;26(2):152-158.
3. Dalgleish AG. Models for HIV pathogenesis. In: Koff WC, Wong-Staal F, Kennedy RC, eds. AIDS Research Reviews. New York, Marcel Dekker, Inc. 1993;(3):73-93.
4. Habeshaw J, Hounsell E, Dalgleish AG. Does the HIV envelope induce a chronic graft-versus-host-like disease? Immunol Today 1992;13(6):207-210.
5. Clerici M, Shearer G, Hounsell EF, Jameson B, Habeshaw J, Dalgleish AG. Alloactivated cytotoxic T cells recognize the carboxy terminal domain of HIV-1gp120 envelope glycoprotein. Eur J Cancer 1993;in press.
6. Lopalco L, De Santis C, Meneveri R, et al. HLA Class I heavy chains are recognized by anti gp120 antibodies in HIV infected individuals. Eur J Immunol 1993;in press.
7. Arthur LO, Bess JW, Sowder RC, et al. Cellular proteins bound to immunodeficiency viruses: Implications for pathogenesis and vaccines. Science 1992;258:1935-1938.
8. Chan WL, Rodgers A, Hancock RD, et al. Protection in simian immuno-

deficiency virus-vaccinated monkeys correlates with anti-HLA Class I antibody response. J Exp Med 1992;176:1203-1207.
9. Abstract Book. IXth International Conference on AIDS. 1993 June 06-11; Berlin. Nos.:A073, A076.
10. Fox CH, Cottler-Fox. The pathobiology of HIV infection. Immunol Today 1992;13(9):353-356.
11. Clerici M, Shearer GM. A Th1→Th2 switch is a critical step in the etiology of HIV infection. Immunol Today 1993;14(3):107-111.
12. De Wit D, Van Mechelen M, Zanin C, et al. Preferential activation of Th2 cells in chronic graft-versus-host reaction. J Immunol 1993;150(2):361-366.
13. Helbert MR, L'age-Stehr J, Mitchison NA. Antigen presentation, loss of immunological memory and AIDS. Immunol Today 1993;14(7):340-344.
14. Moreira AL, Sampaio EP, Zmuidzinas A, Frindt P, Smith KA, Kaplan G. Thalidomide exerts its inhibitory action on tumor necrosis factor by enhancing mRNA degradation. J Exp Med 1993;177:1675-1680.
15. Liu Z, Sun Y-K, Xi Y-P, et al. Contribution of direct and indirect recognition pathways to T cell alloreactivity. J Exp Med 1993;177:1643-1650.

IDIOTYPIC INDUCTION OF AUTOIMMUNE DISEASES: A NEW ASPECT OF INFECTION AND AUTOIMMUNITY

Yehuda Shoenfeld
Research Unit of Autoimmune Diseases
Department of Medicine 'B'
Sheba Medical Center
Tel-Hashomer, 52621
ISRAEL

Introduction

In 1974, Jerne [1] proposed that the immune response might be regulated via the unique antigenic determinants of immunoglobulin variable regions (idiotypes) described earlier by Kunkel et al. [2] and Oudin and Michel [3]. This hypothesis predicts that the idiotypic determinants of each antibody molecule would be complemented by those of another, creating an idiotypic network through which immunoglobulin expression might be controlled. A multitude of experiments has followed in which manipulation of the immune response was achieved through the use of anti-idiotypic reagents [4]. These studies have generally strengthened this hypothesis.

Jerne extended his ideas in 1982 by grouping Ab2s into two functional subsets: those that recognize determinants present in the V region and which do not involve the combining site for the eliciting antigen, and those that represent internal images of the eliciting antigen [5]. He also proposed that naturally occurring Ab2s exist which represent internal images of exogenous antigens, and that these antibodies are important in the establishment of the B cell repertoire. Thus, the idiotypic network might not only be functionally important in the regulation of immune responses, but may also play a role in the establishment of the preimmune repertoire.

According to Jerne the antigen (autoantigen) may stimulate the generation of Ab_1, yet the serologically unique structure of its antigen binding site (Ab_1-idiotype) by itself stimulates the immune system to produce Ab_2. The later recognizing the antigen (anti-Ab_1 or anti-id) binding site of Ab_1, may molecularly mimic the AG either conformationally or even by its amino-acid sequence [6]. If the primary antigen is by itself an antibody (or an autoantibody) we may end with Ab_2 (anti-id) which may be almost identical to the original immunizing autoantibody.

A. G. Dalgleish et al. (eds.), The Impact of Biotechnology on Autoimmunity, 73–80.

Alternatively, Ab_2 may be a regulatory anti-id which may significantly affect the immune reaction either enhancing or abrogating it [7].

Jerne referred to the self regulation of an immune reaction by the sequential cascade of idiotypes and anti-idiotypes. Furthermore, the network was even harnessed for preventive immunization as well as for therapeutic purposes [4,8-10]. Yet, it was proposed that the idiotypic network may be involved in pathogenesis mainly of autoimmune diseases.

Autoimmune diseases may be classified into two groups:

1) Classical autoimmune diseases, fulfilling Witebski-Rose criteria for autoimmunity [11,12].
2) Other autoimmune diseases which are regarded as classical ones, but do not fulfill the above criteria.

The first group entails diseases such as myasthenia gravis, pemphigus vulgaris, autoimmune thrombocytopenia or Graves' thyrotoxicosis. In such diseases, the autoantigen is well defined, the binding of autoantibody to the autoantigen was well documented and the pathogenic mechanisms leading to the phenotypic expression of clinical manifestations are well explained. Pemphigus vulgaris is a severe and often fatal blistering disease, affecting the skin and mucous membranes, in which intraepithelial vesicles form as a result of a distinctive process of cell detachment termed acantholysis. The immunologic abnormalities discovered three decades ago are consistently used as markers of the disease. These investigators showed by immunofluorescence techniques that the majority of patients with pemphigus have IgG-class autoantibodies directed against squamous-epithelial-cell surfaces. Recently they were shown to belong to the cadherin family of Ca^{++} dependent cell adhesion molecules having 130 KD MW [13]. Cutaneous blisters and erosions with the histologic, ultrastructural, and immunofluorescence features of pemphigus occurred in 39 of 55 mice give intraperitoneal injections of IgG from patients with pemphigus and in none of 58 control mice given normal human IgG. IgG fractions with high titers of pemphigus antibodies were most effective in inducing disease, and this effect was dose dependent. Titers of circulating IgG in mouse serum closely correlated with the extent of disease induced [13]. The specific binding of these transfused autoantibodies to the 130 KD autoantigens was also confirmed [13].

The second group of the diseases entails conditions such as systemic lupus erythematosus (SLE), primary biliary cirrhosis (PBC), polymyositis or scleroderma (Sc). All these conditions are suspected as being classical autoimmune disease on the basis of genetic (i.e. HLA-association) hormonal (i.e. female predominance), immunological (the presence of autoantibodies) and environmental (i.e. viruses) reasons [11,12]. In all of them autoantigens are implicated; however, they are either intracellular molecules or/and ubiquitous compounds [e.g. DNA, pyruvate dehydrogenase-(PDH), histidyltransferase (Jo-7) or topoisomerase-I-(topo-I)]. In none of them the immunization of naive animals with the autoantibodies led to the induction of a disease [14,15], nor passive transfusion of the autoantibodies resulted in overt clinical manifestations [16]. Although, some of the autoantibodies were found to have functional characteristics (e.g. inhibiting the autoantigen enzyme), yet nobody can explain the clinical manifestations by these functional capabilities.

Therefore, it seems that despite the specificity of the respective autoantibodies to the diseases, other mechanisms are involved both at the initiation stage of the disease as well as in determining the phenotypic expression of the manifestations.

In the following, we would like to suggest that in the latter group of diseases, idiotypic dysregulation may play a significant role in the pathogenesis, fitting with the recent theory of the immunological humunculus [17,18].

This will be confirmed by different experimental autoimmune models.

The System

Immunization of naive mice with a specific autoantibody to a non-immunogenic (e.g. dsDNA, cardiolipin) or to a nonpathogenic antigen (e.g. pyruvate dehydrogenase enzyme, proteinase-3 enzyme), leads to the generation of Ab_1 - namely an anti-autoantibody (i.e. anti-id). However, if we will follow-up the mice for a longer time (3-5 months), we will notice the *de novo* generation, by the mouse, of Ab_2 (anti-anti-autoantibody), which may simulate the original autoantibody (human or mouse origin) in its binding characteristics. Thus, we end with naive mice producing specific autoantibodies. Not surprisingly this phenomenon is associated with the emergence of the full blown serological, immunohistochemical, and clinical manifestations of the respective disease.

SYSTEMIC LUPUS ERYTHEMATOSUS (SLE)

In a series of experiments we have shown that immunization of various strains of mice [19] with monoclonal [21] or polyclonal [22], human [20,21] or mouse [21] anti-DNA antibodies, mainly carrying a "pathogenic idiotype" ended in [23] the production of a panoply of SLE related autoantibodies by the mouse (anti-DNA, anti-Sm, Anti-Ro/SS-A, anti-histones). The serological markers were associated with typical findings for SLE such as increased sedimentation rate, leukopenia, thrombocytopenia, proteinuria, alopecia and paralysis [22,23]. The deposition of the mouse anti-DNA antibodies in the glomeruli of the kidneys [20], in the skin [22] and the brain was well established. Other groups have confirmed these results [22,24].

Furthermore, a similar scenario of immunization cascade was shown employing "parallel" monoclonal autoantibodies generated by the hybridoma technique, using the lymphocytes of the specific disease induced mice (anti-DNA, anti-La, anti-cardiolipin [25,26].

ANTI-PHOSPHOLIPID SYNDROME (APLS)

The APLS syndrome characterized by recurrent thromboembolic phenomena, recurrent fetal loss and thrombocytopenia [27] is tightly associated with the presence of anti-cardiolipin antibodies or/and lupus anticoagulant [28].

Immunization of various strains of naive mice with human or mouse [29], monoclonal [29] or polyclonal [30,31] anti-cardiolipin antibodies, was followed after 3-4 months by the generation of anti-cardiolipin antibodies by the mice. The mice also exhibited thrombocytopenia [29-31] and their activated thromboplastin time (aPTT) was prolonged. When mated, the immunized female mice had low fecundity and an increased rate of fetal resorptions (the equivalent of human fetal loss). The live fetuses and their compatible placentae were smaller in comparison to those of nonrelevant Ig immunized mice. In some of the fetuses severe thrombosis could be demonstrated [29].

Immunizing mice with a special human anti-DNA antibody (MiV-7) led to the induction of APLS secondary to [26] an association reported in 40% of patients with APLS (e.g. "secondary APLS").

PRIMARY BILIARY DISEASE (PBC)

PBC is a disease seen mainly in women at the fifth and sixth decade and is characterized by a progressive destruction of the bile canaliculi. The latter leads to obstructive jaundice, cirrhosis and severe agonizing pruritus. In early stages, when diagnosed, a pericanalicular T-lymphocytes infiltration can be found [32]. In 95% of the patients one can detect high titers of anti-mitochondrial antibodies [33,34]. These anti-pyruvate dehydrogenase complex autoantibodies characteristically bind to 4 polypeptides shown in immunoblot to have 74, 56, 68, 36 kd molecular weight.

Immunizing BALB/C mice with PDH purified serum autoantibodies from a patient with PBC, resulted after 5 months, in production of anti-PDH antibodies by the mice. (We took care to exclude PDH from the immunizing immunoglobulin.) These antibodies inhibited the enzyme pyruvate dehydrogenase. Six months after the injection of the anti-PDH antibodies we could demonstrate a specific peri-bile canalicular lymphocytic infiltration in the mice livers.

WEGENER'S GRANULOMATOSIS (WG)

This vasculitic condition affecting females and males is characterized among other manifestations by mideline granulomas affecting the skull sinuses, the lungs and the kidneys. The disease starts with leukocytes infiltration with microabscesses, than as a second step one can detect infiltration of T lymphocytes ending in granulomas formation. The disease is associated with a highly specific anti-neutrophil cytoplasmic antibodies (ANCA), which in the case of WG were found to bind mainly to an enzyme, proteinase-3 (29kd) [35].

Immunizing BALB/C mice with purified ANCA antibodies from two patients with active WG, led in the first experiment to the death of the mice from multiple lung microabscesses. In the second experiment, in a longer follow up, we could demonstrate the appearance of granulomas. In both conditions ANCA antibodies confirmed by immunofluorescence assay and Elisa were demonstrated.

Conclusion

In these four experiments we have shown the ability to induce four classical autoimmune conditions: SLE, APLS, PBC and WG following immunization of naive healthy mice in the footpads with the specific autoantibody emulsified in Freund's adjuvant. In all cases the mice developed the specific autoantibody (Ab2) by themselves. We are aware that this induction can not be effected without the involvement of the adjuvant. Jerne reasoned that the great diversity of idiotypes would, to a considerable extent, mirror the diversity of antigenic shapes in the external world. Thus it is conceivable that an external antigen mimicking "pathogenic" or "regulatory" idiotypes may stimulate an increase in its titers therefore triggering the induction of the disease. It seems that molecular mimicry not necessarily is limited to autoantigens, but is expanded also to the idiotypic cascade.

Our results support I. Cohen "Immunological humunculus" detailed recently [17,18]. We postulate that the "natural" analogous to our experimental models resides in antibacterial antibodies carrying the pathogenic idiotypes. Indeed previously we have already reported on the presence of the 16/6 Id pathogenic anti-DNA antibody in sera of patients infected with mycobacteria (pulmonary tuberculosis) [36] and Klebsiella (pneumonia and urinary tract infections) [37,38]. The close relationship between infection and autoimmunity was summarized by us recently [38]. Thus it is conceivable that infection may trigger autoimmune diseases, by inducing antibacterial antibodies carrying the pathogenic idiotypes of autoantibodies (Ab_1). In the presence of the adjuvant effect attributed to various bacteria, these antibodies (Ab_1) start the cascade of idiotypic dysregulation shown by us in the experimental models, leading eventually to the generation of Ab_3 which may either by itself or via regulation bring to an overt clinical autoimmune condition.

In conclusion it seems that antibodies induced by regular bacteria, may carry "pathogenic" idiotypes [32,37,38], thus initiating autoimmune diseases in the presence of the bacteria (adjuvant). In many occasion the disease emerges many months (or years) following the infection and therefore the relationship to the infection is unnoticed.

Acknowledgements

We would like to thank Ms. Shira Lancry for her excellent secretarial assistance.

The studies were supported by the USA-Israel binational fund, a grant given by the Israeli Academy for Sciences and Humanities, grants given by the chief scientist of the Israeli Ministry of Health and the Stanley Burton fund for the research in autoimmunity.

References

1. Jerne NK. Towards a network theory of the immune system. Ann Immunol (Paris) 1974;125c:373-389.
2. Kunkel HG, Mannik M, Williams RC. Individual antigenic specificity of isolated antibodies. Science 1963;140:1218-1219.
3. Oudin J, Michel M. Une nouvelle forme d'allotyle des globulines je du serum de lapin apparement liee a le fonction et a la specificite anticorps C.R. Seances Soc Biol Paris 1963;257:805-808.
4. Cazenave PA (ed.). Anti-idiotypic vaccines in progress in vaccinology. Springer Verlag, New York, 1991 pp. 1-138.
5. Jerne NK, Roland J, Casenave PA. Recurrent idiotopes and internal images. Embo J 1982;1:243-247.
6. Bruck C, Man SC, Slaoul M, Gaulton G, Smith T, Fields B, Mullins J, Green M. Nucleic acid sequence of an internal image bearing monoclonal anti-idiotype and its comparison to the sequence of the external antigen. Proc Nat Acad Sci (USA) 1986;83:6578-6582.
7. Bona CA. Regulatory idiotypes In: Modern Concepts in Immunology II. John Wiley Sons, New York, 1987 pp. 1-127.
8. Smordinsky NI, Ghendler Y, Bakimer F, Chaitchuk S, Keydar I, Shoenfeld Y. Towards an idiotype vaccine against mammary tumors. Induction of an immune response to breast concer-associated antigens by anti-idiotypic antibodies. Eur J Immunol 1989;18:1713-1718.
9. Carson DA, Chen PP, Kipps TJ (eds.). Idiotypes in Biology and Medicine, 1990. Karger, Basel, Switzerland pp. 1-209.
10. Buskila D, Shoenfeld Y. Manipulation of anti-DNA idiotypes. A possible treatment approach to autoimmune diseases. Clin Molec Aspects of Autoimmune Dis (Karger) 1992;8:114-129.
11. Shoenfeld Y, Isenberg DA. The Mosaic of Autoimmunity. Elsevier, Amsterdam, 1988 pp. 1-29.
12. Shoenfeld Y. Autoimmune diseases: multiple factors involved in the etiology. Isr J Med Sci 1988;24:351-352.
13. Amagai M, Karpati S, Prussick P, Klau-Kovtum V, Stanley JR. Autoantibodies against amino-terminal cadherin-like binding domain of pemphigus vulgaris antigen are pathogenic. J Clin Invest 1992;90:919-926.
14. Madaio MP, Hodder S, Schwartz RS, Stollar BD. Responsiveness of autoimmune and normal mice to nucleic acid antigens. J Immunol 1984;132:872-882.

15. Edlerg JC, Tosic L, Taylor RP. Immune adherence and the processing of soluble complement fixing antibody DNA immune complexes in mice. Clin Immunol Immunopathol 1989;51:118-132.
16. Krams SM, Surh CD, Coppel RL, Ansar A, Ruebner B, Gershwin ME. Immunization of experimental animals with dihydrolipoamide acetyltransferase, as a purified recombinant polypeptide, generates mitochondrial antibodies but not primary biliary cirrhosis. Hepatology 1989;9(3):411-416.
17. Cohen IR. The cognitive principle challange clonal section. Immunol Today 1992;13:441-446.
18. Cohen IR. The cognitive paradigm and the immunological humunculus. Immunol Today 1992;490-494.
19. Mendlovic S, Brocke S, Fricke H, Shoenfeld Y, Bakimer R, Mozes E. The genetic regulation of the induction of experimental SLE. Immunology 1990;69:228-236.
20. Mendlovic S, Brocke S, Shoenfeld Y, Ben-Bassat M, Meshorer A, Bakimer R, Mozes E. Induction of a SLE-like disease in mice by a common anti-DNA idiotype. Proc Natl Acad Sci (USA) 1988;85:2260-2264.
21. Blank M, Krup M, Mendlovic S, Fricke H, Mozes E, Talal N, Coates ARM, Shoenfeld Y. The importance of the pathogenic 16/6 idiotype in the induction of SLE in naive mice. Scand J Immunol 1990;31:45-52.
22. Tincani A, Balesterieri G, Allegri F, Cattaneo R, Fornasieri A, Sinico A, D'Amico G. Induction of experimental SLE in naive mice by immunization with human polyclonal anti-DNA antibody carrying the 16/6 idiotype. Clin Exp Rheumatol 1993; in press.
23. Shoenfeld Y, Mozes E. Pathogenic idiotypes of antibodies and autoimmunity: A lesson from a new syndrome of experimental SLE. FASEB J 1990;4:2646-2651.
24. Rombach E, Stetler DA, Brown JC. Rabbits produce SLE-like anti-RNA polymerase I and anti-DNA autoantibodies in response to immunizationwith either human or murine SLE anti-DNA antibodies. Autoimmunity 1992;13:291-302.
25. Fricke H, Offen D, Mendlovic S, Shoenfeld Y, Bakimer R, Sperling J, Mozes E. Induction of experimental SLE in mice by immunization with a monoclonal antiLa autoantibody. Int Immunol 1989;2:225-230.
26. Blank M, Krause I, Ben-Bassat M, Shoenfeld Y. Induction of experimental anti-phospholipid syndrome associated with SLE following immunizationwith human monoclonal pathogenic anti-DNA idiotype. J Autoimmunity 1992;5:495-509.
27. Tomer Y, Cohen J, Shoenfeld Y. Clinical aspects of antiphospholipid antibody syndrome. Harefuah 1991;120:514-520.
28. Sukenik S, El-Roiey A, Shoenfeld Y. Lupus anticoagulant and anticardiolipin antibodies in SLE. Acta Hematol 1986;76:86-89.

29. Bakimer R, Fishman P, Blank M, Sredni B, Djaldetti M, Shoenfeld Y. Induction of primary antiphospholipid syndrome in mice by immunization with a hyman monoclonal anti-cardiolipin antibody (H-3). J Clin Invest 1992;89:1558-1663.
30. Cohen J, Bakimer R, Blank M, Valesini B, Shoenfeld Y. Anti-phospholipid syndrome induced by active immunization with serum IgG anti-cardiolipin antibodies (submitted).
31. Blank M, Tincani A, Shoenfeld Y. Induction of experimental anti-phospholipid syndrome in naive mice with purified IgG anti-phosphatidylserine antibodies. J Rheumatol, in press.
32. Weiss P, Shoenfeld Y. Primary biliary cirrhosis: Increasing problem. Isr J Med Sci 1992;28:726-728.
33. Zurgil N, Bakimer R, Kaplan N, Youinou P, Shoenfeld Y. Anti-pyruvate dehydrogenase in primary biliary cirrhosis. J Clin Immunol 1991;11:239-245.
34. Zurgil N, Bakimer R, Moutsopolous HM, Youinno P, Kaplan MM, Shoenfeld Y. Antimitochondrial (pyruvate dehydrogenase) autoantibodies in autoimmune rheumatic diseases. J Clin Immunol 1992;12(3):201-209.
35. Kallenberg C, Munthr HL, Cohen JW. ANCA: a still growing class of autoantibodies in inflamatory disord (Review). Amer J Med 1992;93:675-682.
36. Sela O, El-Rociy A, Isenberg DA, Kennedy RC, Colaco CB, Pinkhas J, Shoenfeld Y. A common anti-DNA antibody idiotype in sera of patients with active pulmonary tuberculosis. Arthritis Rheumat 1987;30:50-55.
37. El-Roeiy A, Sela O, Isenberg DA, Kennedy RL, Shoenfeld Y. The sera of patients with Klebsiella infections contain a common anti-DNA idiotype (16/6 Id) and anti-polynucleotide activity. Clin Exp Immunol 1987;67:507-515.
38. Shoenfeld Y, Cohen IR. Infection and autoimmunity. In: The Antigens, M. Sela (ed.) Academic Press, 1987, pp. 307-325.

IMMUNOPATHOLOGY OF INSULIN-DEPENDENT DIABETES MELLITUS (IDDM)

Jean-François Bach
Inserm U25
Hôpital Necker
Paris
FRANCE

Problems of Definition

Type I diabetes is defined by insulin requirement as opposed to type II diabetes where insulin therapy is not regularly required. It is now apparent that this clinically very useful classification will evolve towards new definitions based on genetic factors and immunological mechanisms. Autoimmunity is considered to be at the origin of the majority of cases of IDDM. In fact, the autoimmune origin involves more cases than conventional type I diabetes: some patients with severe type II diabetes (they may require transient or durable insulin therapy) share with conventional type I diabetes HLA DR3 or DR4 genes and the presence of islet cell antibodies (slow Type 1 diabetes).

Diabetes Prediction

Present research in the immunology of IDDM is directed towards the immunologic prevention of the disease. This goal which may eventually lead to the eradication of the disease implies that the pathogenic process is recognized early enough, before insulin is fully required since a large fraction of insulin-producing β cells is then ineluctably destroyed. This early diagnosis can rely on detection of genetic predisposition factors but this genetic approach is not sufficient because the penetrance of the predisposing genes is only modest: no more than 40% monozygotic twins are concordant for Type 1 diabetes. Thus, genetic studies must be complemented by epidemiologic studies searching for the environmental factors that modulate the disease frequency and immunological studies attempting to detect the onset of the pathogenic process before the emergence of clinical diabetes. When this early disease recognition is achieved, one may envision the application of an efficacious and innocuous therapy that will stop the pathologic process before irreversible lesions have developed.

A. G. Dalgleish et al. (eds.), The Impact of Biotechnology on Autoimmunity, 81–85.

Predisposition Genes

HLA GENES

Converging arguments indicate that HLA genes play the central role in predisposition to IDDM. There is a strong association of the disease with HLA DR3 and DR4 genes. This association has now been extended to DQ alleles [1,2] and more particularly to certain residues, notably residue 57 on the HLA DQβ chain (Asp is found with particularly low frequency at this position in IDDM) [3]. Similar data are found in the nonobese diabetic (NOD) mouse, an experimental model of human IDDM. It has been shown in these mice from genetic segregation studies using various crosses of NOD mice with nondiabetic strains that the H-2 region (I-A genes) accounted for the strongest genetic predisposition and that the change in the H-2 genes by use of transgenesis prevented the onset of IDDM [4]. Interestingly, there is no I-E expression in NOD mice and I-E transgenic NOD mice are protected from diabetes [5], which suggests that this anomaly could contribute to the defective T cell mediated suppression shown by these mice.

MHC NON-HLA GENES

The absolute IDDM risk conferred by HLA genes to the general population is of the order of 1-2% which is a figure much lower than the disease concordance rate observed between HLA identical siblings (10-12%). One may hypothesize that non-HLA genes located within the major histocompatibility complex (MHC), that are cotransmitted with HLA genes within families but not in the general population explain this discrepancy. Several genes of this category are now known, notably the genes coding for proteasomes (LMP2) and for peptide transporters (TAP). Several groups have recently demonstrated a strong negative association between a TAP2 allele (0201) and IDDM [6,7] suggesting that the presence of this allele is protective from the disease. It remains to be demonstrated whether this association is explained or not by a linkage disequilibrium with class II alleles: there is evidence that two extended haplotypes include both class II and TAP2 alleles but this linkage disequilibrium does not apply to all haplotypes. Additionally, we have observed that TAP2 and class II associations are cumulative [7].

NON-MHC GENES

The difference between disease concordance rates in HLA identical siblings (~ 12%) and monozygotic twins (~ 35%) shows the major role of non-HLA genes in predisposition to IDDM. Only one such gene has been identified in human IDDM, in the region of the insulin gene (perhaps the insulin gene itself) [8]. Other non-MHC nongenes have been recognized in the NOD mouse such as the interleukin 2 gene and a gene in the bc12 locus or in its vicinity [9,10]. Bc12 is a proto-oncogene controlling lymphoid cell apoptosis. Taken together with observations made in MRL/1 mice where the lpr trait has been identified to a mutation in the fas gene that induces abnormal survival of T cells, one may assume that bc12 contributes to abnormal survival of islet cell autoreactive T cells. Segregation studies in NOD mice have indicated a role for other non-MHC genes [8] but it is yet difficult to establish the weight of these various genes in IDDM pathogenic and potential usefulness for disease prediction.

Environmental Factors

The importance of environmental factors in human IDDM is suggested by the low penetrance of predisposing genes mentioned above. Only limited data are available to define these factors. Much interest has been given recently to the possible triggering role of cow milk during infancy [11]. There appears to exist in bovin milk a 69 kDa protein that cross reacts with an islet cell autoantigen that could stimulate an autoimmune T cell response via a molecular mimicry mechanism. The search for diabetogenic viruses is another approach that has proven difficult so far. In the animal models, emphasis has recently been put on the protective influence of a number of infectious agents such as viruses (LCMV, LDHIV) or bacteria. The complete protection afforded by Freund's complete adjuvant is particularly striking [12].

Immunological Predictive Tests

A number of autoantibodies are evidenced in the serum of prediabetics: islet cell antibodies (immunofluorescence), anti-insulin and antiproinsulin antibodies, antibodies against glutamic acid decarboxylase (GAD, anti-69 kDa protein). Some of these antigens give rise to T cell reactivity which is important in view of the presumed major responsibility of T cells in IDDM pathogenesis.

Immunotherapy

Randomized therapeutic trials using cyclosporin and azathioprine have convincingly shown that chemical immunosuppression may induce durable remission of insulin dependency when applied within 6 weeks after initiation of insulin therapy [13]. The effect is not permanent, however, and one is obviously concerned with the necessity of administering chronic immunosuppressive therapy in such young patients.

Two directions are now taken to circumvent these difficulties with the goal of applying them at earlier stages of the disease as can be identified by the genetic and immunologic tests described above.

The first approach is based on environmental factors: avoidance of cow milk, stimulation with substitutes of innocuous infectious agents such as BCG.

The second one aims at restoring the lost tolerance to self islet autoantigens. This may be directly attempted by administering islet cell autoantigens (insulin, GAD, etc.) eventually by the oral route, which has recently shown to be highly tolerogenic [14]. Alternatively, one may think of using the islet cell autoantigens persisting in situ, under the cover of monoclonal anti-T cell or anti-adhesion monoclonal antibodies. This latter approach has recently provided remarkable results in the NOD mouse when long lasting remission of established diabetes was obtained with an anti-CD3 monoclonal antibody [15]. These approaches could be helped by clearing the insulitis by short term therapy with cell depleting agent such as IL-2 toxin that selectively depletes activated T cells.

Conclusion

It is the combination of these approaches that will ultimately allow the prevention of IDDM. It should be realized though that a major factor of complication derives from the fact that each of the two basic approaches, prediction and immunotherapy, depends on the other one: one cannot easily generate prediction programs without the hope for immunotherapy and one cannot apply immunotherapy successfully without early diagnosis. This apparent paradox prompts the undertaking of the two approaches in parallel with the clear assumption that the final solution (prevention of IDDM) will require a few years of intense clinical research before ensuring direct benefit to patients.

References

1. Baisch JM, Weeks T, Giles R, Hoover M, Stastny P, Capra JD. Analysis of HLA-DQ genotypes and susceptibility in insulin-dependent diabetes mellitus. New Engl J Med 1990;322:1836-41.
2. Caillat-Zucman S, Garchon HJ, Timsit J, Assan R, Boitard C, Djilali-Saiah I, Bougnères P, Bach JF. Age-dependent HLA genetic heterogeneity of type 1 insulin-dependent diabetes mellitus. J Clin Invest 1992;90:2242-50.
3. Todd JA, Bell JI, McDevitt HO. HLA-DQβ gene contributes to susceptibility and resistance to insulin-dependent diabetes mellitus. Nature 1987;329:599-604.
4. Slattery RM, Kjer-Nielsen L, Allison J, Charlton B, Mandel TE, Miller JFAP. Prevention of diabetes in non-obese diabetic I-Ak trangenic mice. Nature 1990;345:724-6.
5. Nishimoto H, Kikutani H, Yamamura KI, Kishimoto T. Prevention of autoimmune insulitis by expression of I-E molecules in NOD mice. Nature 1987;328:432-4.
6. Ronningen KS, Undlien DE, Plotski R, Maouni N, Konrad RJ, Jensen E, Hornes E, Reijonen H, Colonna M, Monos DS, Strominger JL, Thorsby E. Linkage disequibrium between TAP2 variants and HLA class II alleles. No

primary association between TAP2 variants and insulin dependent diabetes mellitus. Eur J Immunol 1993;23:1050-6.

7. Caillat-Zucman S, Bertin T, Timsit J, Boitard C, Assan R, Bach JF. Protection from insulin dependent diabetes mellitus to a peptide transporter gene. Eur J Immunol 1993;23:1748-8.
8. Julier C, Hyer RN, Davies J, Merlin F, Soularue P, Briant L, Cathelineau G, Deschamps I, Rotter JI, Froguel P, Boitard C, Bell JI, Lathrop GM. Insulin-IGF2 region on chromosome 11p encodes a gene implicated in HLA-DR4-dependent diabetes susceptibility. Nature 1991;354:155-9.
9. Todd JA, Aitman TJ, Cornall RJ, Ghosh S, Hall JRS, Hearne CM, Knight AM, Love JM, McAlleer MA, Prins JB, Rodrigues N, Lathrop M, Pressey A, DeLarato NH, Peterson LB, Wicker LS. Genetic analysis of autoimmune type 1 diabetes mellitus in mice. Nature 1991;351:542-7.
10. Garchon HJ, Bedossa P, Eloy L, Bach JF. Identification and mapping to chromosome 1 of a susceptibility locus for periinsulitis in non-obese diabetic mice. Nature 1991; 260-2.
11. Karjalainen J, Martin JM, Knip M, Ilonen J, Robinson BH, Savilahti E, Âkerblom HK, Dosch HM. A bovine albumin peptide as a possible trigger of insulin-dependent diabetes mellitus. New Engl J Med 1992;30:302-7.
12. McInerney MF, Pek SB, Thomas DW. Prevention of insulitis and diabetes onset by treatment with complete Freund's Adjuvant in NOD mice. Diabetes 1991;40:715-25.
13. Feutren G, Papoz L, Assan R, Vialettes B, Karsenty G, Vexiau P, Du Rostu H, Rodier M, Sirmai J, Lallemand A, Bach JF. Cyclosporin increases the rate and length of remission in insulin-dependent diabetes of recent onset. Results of a multicentre double-blind trial. Lancet 1986;2:119-24.
14. Weiner HL, Miller A, Khoury SJ, Zhang ZJ, Al-Sabbagh A, Brod SA, Lider O, Higgins P, Sobel R, Matsui M, Sayegh M, Carpenter C, Eisenbarth G, Nussenblatt RB, Hafler DA. Suppression of organ-specific autoimmune diseases by oral administration of autoantigens. Progress in Immunology VIII. In: Proceeding of the 8th International Congress of Immunology. Budapest 1992, J. Gergely et al. (Eds.), Springer Hungarica 1993, pp. 627-634.
15. Chatenoud L, Thervet E, Primo J, Bach JF. Anti-CD3 antibody induces long-term remission of overt autoimmunity in non-obese diabetic mice. Proc Natl Acad Sci USA, in press.

AUTOIMMUNE MECHANISMS IN RHEUMATOID SYNOVIAL INFLAMMATION

Ingrid Randen, Keith M. Thompson, Øystein Førre*, and Jacob B. Natvig
Institute of Immunology and Rheumatology and
**Oslo Rheumatism Hospital*
NORWAY

Rheumatoid arthritis is a chronic, autoimmune disease characterized by early macrophage and lymphocyte infiltration of the synovial membrane. As the disease progresses there is an increased vascularization, and T- and B-cells are found around the blood vessels. Surrounding these cell clusters, sometimes described as germinal center-like, there is a mixture of cells, including undifferentiated blast cells, plasma cells, dendritic cells and macrophages. Due to activation and communication among these cells there is an unrestrained proliferation of macrophages, T- and B-lymphocytes, and production of cytokines and immunoglobulins. The synovial inflammation eventually destroys the joint.

In the synovium T-cells predominate over the B-cells and CD4+ cells predominate over CD8+ cells. B cells are located primarily within the lympoid cell clusters and the overall arrangement is consistent with T-dependent B-cell activation.

In an attempt to shed some light over two of the main actors in the synovial inflammation, we have cloned and characterized B- and T-lymphocytes from the synovial tissue of patients with rheumatoid arthritis [1]. The variable regions of the respective immunoglobulins and T-cell receptors have been functionally characterized and sequenced at the nucleotide level [2-4]. In this brief summary we will focus on the B cells in the rheumatoid inflammation.

In rheumatoid arthritis (RA) there is a strong link between disease and the serological presence of antibodies binding the Fc region of IgG molecules, so called rheumatoid factor (RF). RF are found in the serum of approximately 80% of patients with RA and the titers often correlate with the severity of the disease. However, RF are particularly abundant in the synovial fluid and tissue of patients with RA [5-7]. Here they form large, complement-forming complexes which are thought to contribute to the sustained inflammation [8,9]. Before the development of hybridoma technology, most of the information about the genetic origin of RF was from diseases like mixed cryoglobulinemia and Waldenströms macroglobulinemia, where there is a monoclonal component in the serum.

A. G. Dalgleish et al. (eds.), The Impact of Biotechnology on Autoimmunity, 87–91.

Immunoglobulins from these diseases often have RF activity although the patient has no symptoms of rheumatoid arthritis.

We have made hybridomas secreting IgM and IgG RF from the synovial tissues of two patients with RA (TS and SJ) and one patient with the polyarticular form of juvenile RA (KL) [10,11]. More than 20 monoclonal RF have been extensively characterized for the fine binding specificities for IgG subclasses and allotypes, showing that the predominant specificity of these synovial tissue derived RF was the Ga specificity, i.e. binding to IgG1, IgG2 and IgG4, but antibodies binding all four human IgG subclasses were also seen, together with an anti-allotypic antibody binding G3m(u) (Table 1).

Variable region gene usage of the IgM RF were tested for binding to V_H and Vκ associated murine monoclonal antibodies and shown to use mainly different variable region genes for their antigen binding site than M-component RF [12]. RF from rheumatoid arthritis patients were largely using heavy chain variable region gene segments belonging to the V_HIII family and a wide variety of light chain gene segments, not restricted to the VκIII as seen among M-component RF [2], indicating that different mechanisms are involved in RF expression in RA compared to RF in B-cell dyscrasias. These findings may prove even more significant as recent investigation of RF from normal individuals produced after exogenous antigen stimulation seem to use variable region gene segments characteristic of M-component RF rather than those used by RF from RA patients [13].

Nucleotide sequencing studies of the genes encoding the antigen binding site revealed that most of the IgM RF were not very mutated [14-16]. Most surprisingly, however, two of the IgM RF were clonally related, i.e. being the progeny of a common ancestor cell. One of them, RF-SJ2, was very close to germline configuration, while the "sister cell", RF-SJ1 had acquired a total of 36 nucleotide changes compared to RF-SJ2. Affinity measurements showed that the extensively mutated RF had a hundred fold higher affinity for human IgG than the RF close to germline [17]. These findings indicate that at least some IgM RF can undergo affinity maturation in the rheumatoid synovial tissue and indicate clonal expansion of certain IgG binding immunoglobulins.

IgG RF are thought to be of particular importance as they have the possibility of self-association and forming complement-fixing complexes [18,19]. Synovial IgG RF from the three patients in our studies formed intra-cellular complement-fixing complexes indicative of a self-association process *in vivo* [11]. Nucleotide sequencing showed that they were in general more mutated than the IgM RF and the mutations led to amino acid substitutions clustered in the hypervariable regions, indicating that they were a product of an antigen driven immune response [20].

Overall these studies of the RF produced in the rheumatoid synovial tissue suggest that the variable region gene usage is diverse and different from that of M-

TABLE 1:CHARACTERISTICS OF SYNOVIAL TISSUE DERIVED RF

Clone	Specificity[1]	Affinity[2] $[M^{-1}]$	V_H	%homology[3]	V_L	%homology
IgM						
RF-SJ2	Ga	3.7×10^5	III	99.3	λI	96.3
RF-SJ1	Pan	4.2×10^7	III	96.3	λI	93
RF-TS5	Ga related	4.8×10^6	III	99.3	κI	97.9
RF-SJ3	Ga	2.2×10^7	III	99.3	κIIIb	98.7
RF-TS2	Ga	1.4×10^7	III	97.3	κIIIa	97
RF-TS4[4]	G3m(u)	ND[5]	III	95.8	κI	96.5
RF-KL1	Ga related	1.5×10^6	III	94.2	κI	97.5
PR-TS1	Poly	ND	III	99.3	λI	98.3
PR-SJ2	Poly	1.8×10^7	III	95.6	λII	90.8
RF-TS1	Pan	1.3×10^7	I	100.0	κIIIb	99.3
RF-TS3	Ga	1.4×10^7	I	99.3	κII	100
RF-SJ4	Ga	1.3×10^7	IV	95.4	κIIIb	100
PR-SJ1	Poly	5.3×10^5	IV	100	λV	83
PR-TS2	Poly	ND	IV	100	λIII	100
IgG						
RF-TS7	Non-a	ND	I	99.3	λII	98.6
RF-KL5	ND	ND	III	96.9	κIV/VI	-
RF-SJ5	ND	ND	III	93.3	κI	94.7

1: The fine specificities of the RF are: Pan: Reaction with all four human IgG subclasses, Ga: Reaction with IgG1, IgG2 and IgG4
Ga related: Reaction with IgG1, IgG2; IgG4 and the G3m(s,t) allotype,
G3m(u): Anti-allotypic specificity; reaction with G3m(b) and G3m(g) on IgG3, Poly: Polyreactive antibody, reacting with various exo- and auto-antigens, Non-a (or nG1m(a)): Reaction with G1m(f), but not G1m(a)
2: The functional affinity constant $K_A=1/K_D$ is deduced from Scatchard plots
3: Homologies are calculated at the nucleotide level after comparison with closest germline or expressed gene
4: Partly heavy chain variable region sequence
5: ND = Not determined (RF-SJ1 and RF-SJ2 are the clonally related antibodies)

component RF. Although most of the IgM RF variable region genes were in or close to germline configuration, evidence that some RF may be a product of an antigen driven immune response was found for the affinity maturated RF-SJ1 and the replacement substitutions in the CDRs of the IgG RF. Some features of the synovial tissue derived RF are summarized in Table 1.

References

1. Randen I, Wilson KB, Thompson KM, Quayle A, Pascual V, Førre Ø, Capra JD, Natvig JB. Autoimmune mechanisms in rheumatoid synovial inflammation. Abstract from meeting on the Impact of Biotechnology on Autoimmunity, Florence, 1993.
2. Randen I, Thompson KM, Pascual V, Victor K, Beale D, Coadwell J, Førre Ø, Capra JD, Natvig JD. Rheumatoid factor V genes from patients with rheumatoid arthritis are diverse and show evidence of an antigen driven response. Immunol Rev 1992;128:49-71.
3. Quayle AJ, Wilson KB, Li SG, Kjeldsen-Kragh J, Oftung F, Shinnick F, Sioud M, Førre Ø, Capra JD, Natvig JB. Peptide recognition, T cell receptor usage and HLA restriction elements of human heat-shock protein (hsp) 60 and mycobacterial 65-kDa hsp-reactive T cell clones from rheumatoid synovial fluid. Eur J Immunol 1992;22:1315-1322.
4. Wilson KB, Quayle AJ, Suleyman S, Kjeldsen-Kragh J, Førre Ø, Natvig JB, Capra JD. Heterogeneity of the TCR repertoire in synovial fluid T lymphocytes responding to BCG in a patient with early rheumatoid arthritis. Scand J Immunol 1993;38:102-112.
5. Mellors RC, Heimer R, Corcos J, Korngold L. Cellular origin of rheumatoid factor. J Exp Med 1959;110:875.
6. Hannestad K. Presence of aggregated γG-globulin in certain rheumatoid synovial effusions. Clin Exp Immunol 1967;2:511-529.
7. Munthe E, Natvig JB. Characterization of IgG complexes in eluates from rheumatoid tissue. Clin Exp Immun 1971;8:149.
8. Munthe E, Natvig JB. Immunoglobulin classes, subclasses and complexes of IgG rheumatoid factor in rheumatoid plasma cells. Clin Exp Immunol 1972;12:55-70.
9. Nardella FA, Gilliland BC, Mannik M. Detection of intermediate complexes by evaluation of the difference between γ-globulin and IgG concentrations. Arthritis Rheumatism 1979;22:141-144.
10. Randen I, Thompson KM, Natvig JB, Førre Ø, Waalen K. Human monoclonal rheumatoid factors derived from the polyclonal repertoire of rheumatoid synovial tissue: production and characterization. Clin Exp Immunol 1989;78:13-18.
11. Randen I, Thompson KM, Thorpe SJ, Førre Ø, Natvig JB. Human monoclonal IgG rheumatoid factors from the synovial tissue of patients with

rheumatoid arthritis. Scand J Immunol 1993;37:668-672.

12. Thompson KM, Randen I, Natvig JB, Mageed RA, Jefferis R, Carson DA, Tighe H, Førre Ø. Human monoclonal rheumatoid factors derived from the polyclonal repertoire of rheumatoid synovial tissue: incidence of cross-reactive idiotopes and expression of VH and V kappa subgroups. Eur J Immunol 1990;20:863-868.
13. Thompson KM, Randen I, Børrezen M, Førre Ø, Natvig JB. V-gene usage of human monoclonal rheumatoid factors derived from healthy donors following immunization. Submitted.
14. Pascual V, Randen I, Thompson K, Sioud M, Førre Ø, Natvig JB, Capra JD. The complete nucleotide sequences of the heavy chain variable regions of six monospecific rheumatoid factors derived from Epstein-Barr virus-transformed B cells isolated from the synovial tissue of patients with rheumatoid arthritis. Further evidence that some autoantibodies are unmutated copies of germ line genes. J Clin Invest 1990;86:1320-1328.
15. Victor KD, Randen I, Thompson K, Førre Ø, Natvig JB, Fu SM, Capra JD. Rheumatoid factors isolated from patients with autoimmunue disorders are derived from germline genes distinct from those encoding the Wa, Po, and Bla cross-reactive idiotypes. J Clin Invest 1991;87:1603-1613.
16. Pascual V, Victor K, Randen I, Thompson KM, Steinitz M, Førre Ø, Fu SM, Natvig JB, Capra JD. Nucleotide sequence analysis of rheumatoid factors and polyreactive antibodies derived from patients with rheumatoid arthritis reveals diverse use of VH and VL gene segments and extensive variability in CDR-3. Scand J Immunol 1992;36:349-362.
17. Randen I, Brown D, Thompson KM, Hughes-Jones N, Pascual V, Victor K, Capra JD, Førre Ø, Natvig NB. Clonally related IgM rheumatoid factors undergo affinity maturation in the rheumatoid synovial tissue. J Immunol 1992;148:3296-3301.
18. Natvig JB, Munthe E. Self-associating IgG rheumatoid factor represents a major response of plasma cells in rheumatoid inflammatory tissue. Ann NY Acad Sci 1975;256:88-95.
19. Mannik M, Nardella FA. IgG rheumatoid factors and self-association of these antibodies. Clinics Rheum Dis 1985;11:551-572.
20. Randen I, Pascual V, Victor K, Thompson KM, Førre Ø, Capra JD, Natvig JB. Synovial IgG rheumatoid factors show evidence of an antigen driven immune response and a shift in the V gene repertoire compared to IgM rheumatoid factors. Eur J Immunol 1993;23:1220-1225.

ANTINUCLEAR ANTIBODIES: CLINICAL AND DIAGNOSTIC RELEVANCE

Laura Caponi, Rossella Neri, Antonio Tavoni, Fortuna Scamardella, Barbara Marchini, Paola Migliorini, and Stefano Bombardieri
Clinical Immunology Unit
University of Pisa
Pisa
ITALY

Introduction

Antibodies directed to intracellular components are the most important serological diagnostic markers for connective tissue diseases (CTD), and may provide useful insights into their pathogenetic mechanisms. Some of these antibodies are considered to be a "marker" of specific diseases while others appear to be linked with specific clinical manifestations. The detection of these autoantibodies in biological samples relies in tests of different specificity, sensibility and simplicity. The development of more sophisticated techniques of detection of autoantigens has lead to the comparison between these methods during international workshops [1].

We will discuss here three aspects of antinuclear antibodies, underlying their importance in clinical medicine:1) their diagnostic relevance, considering not only the specific "marker" autoantibodies, but also antibody associations; 2) additional information which can be obtained by studying the more fine specificities of these antibodies; 3) the preliminary evidences that in certain instances the study of the intimate structure of the antigens can provide important clues in the etiopathogenesis of the CTD.

Antinuclear Antibodies as Diagnostic "Markers"

In order to study the diagnostic role of antinuclear antibodies, 1457 sera from 615 patients with various CTD, followed at the Clinical Immunology, Rheumatology Unit of the University of Pisa, were tested.

Antinuclear antibodies, anti-dsDNA antibodies and anticentromere antibodies were detected by means of indirect immunofluorescence, respectively with rat liver, *Crithidia Luciliae* or Hep2 cells as substrate. Counterimmunelectrophoresis (CIE) recently validated during a European workshop [2], using either commercial thymus extract or human spleen as substrate, was performed to detect other specificities,

A. G. Dalgleish et al. (eds.), The Impact of Biotechnology on Autoimmunity, 93–97.

i.e. Ro/SSA, La/SSB, Sm, RNP, Scl70, PCNA, Sl, Jo1. Only patients meeting well established criteria for the diagnosis of various CTD were included in the study. Patients included 155 with systemic lupus erythematosus (SLE), 78 with primary Sjøgren's syndrome (1°SS), 18 with mixed connective tissue disease, 23 with idiopathic Raynaud's phenomenon, 87 with rheumatoid arthritis, 33 with unclassified connective tissue disease (UCTD), 9 with polymyositis (PM), 54 with mixed cryoglobulinemia, and 157 with systemic sclerosis (SSc).

Table 1. Antinuclear antibodies associations

a) Disease "Markers"

SLE	PCNA	8.0%
	Sm	12.0%
	Sl	11.0%
	dsDNA	12.0%
	> 4 specificities	3.5%
PM	Jo1	20.0%
SSc	ACA	18.0%
	Scl70	41.0%

b) Antibody specificities or antibody associations indicative of disease states.

Ro/SSA and La/SSB	1°SS, SLE, UCTD, others
RNP alone	MCTD, SCE, 1°SS, UCTD, others
more than 3 specificities	SLE, UCTD, others

c) Antibody specificities or antibody associations not specific of disease states.

snRNP
Ro/SSA
snRNP with Ro/SSA

In Table 1 are summarized antibody associations. It appears clear that an antibody profile may provide sometimes much broader diagnostic information than the evaluation of single marker autoantibodies.

Clinical Association of Finer Specificities in Antinuclear Antibodies

It is now well known that the same autoantibody may be present in different CTD. On the other hand it has been clearly demonstrated that complex nuclear antigens, contain several antigenic determinants. It has been postulated that the determination of major epitopes recognized by naturally occurring autoantibodies in CTD may help to increase their clinical relevance and to identify a sort of disease specific epitopes. Preliminary studies seem to indicate that the identification of finer specificities, i.e. the epitopes involved, could be even more important than the presence/absence of a specific antibody in a clinical/serological relationship for the identification of pathogenetic autoantibodies.

For instance, it is well known that anti-Ro/SSA antibodies are directed on immunoblotting (IB) toward two proteinaceous components: one of 52 KD and the other of 62 KD. A number of studies shows that passive transfer of these antibodies may cause the so-called "neonatal lupus" characterized by congenital complete hearth block. Recent studies demonstrated that antibodies directed against the 52 KD protein rather than those directed against the 60 KD are responsible of this event [3].

We examined clinical associations of the two subunits of Ro/SSA antigen, using a biochemically purified antigen isolate from human spleen [4]. IB analysis was performed on 34 sera of primary Sjøgren syndrome (1°SS) patients and on 35 sera of SLE patients with anti-Ro/SSA antibodies previously detected with CIE. While SLE sera reacted preferentially with 60 Kd Ro/SSA alone (46% vs 29%), 1°SS sera reacted almost equally to 52 Kd Ro/SSA in addition to 60 Kd polypeptide (41% vs 37%) [5].

Similarly, we tested 26 SLE patients and 42 patients with 1°SS with anti-La/SSB autoantibodies detected by CIE. Using a semipurified antigen from rabbit thymus. The La/SSB antigen separates itself in 4 polypeptides of 46 KD, 30 KD, 24 KD, and 20 KD of molecular weight [6]. In Table 2 we summarized the results.

Table 2. Reactivity of SLE and 1°SS sera with La/SSB antigen.

Reactivity to	SLE	1°SS
46 KD, 30 KD, 24 KD, and 20 KD	19/26 (74%)	20/42 (47%)
46 KD and 20 KD	4/26 (15%)	10/42 (23%)
46 KD	3/26 (11%)	12/42 (30%)

As shown there are clear differences between SLE and 1°SS in their reactivity with different epitopes of La/SSB antigen. At present a study is under way to verify if this kind of autoantibodies (or a subset among them) correlates with a high focus score in biopsies of salivary glands.

Autoantibodies against ribosomal proteins appear to be rather specific toLE patients. The relationship between the detection of these antibodies and clinical manifestations is at the moment controversial. A number of studies reported correlations between their presence and psychosis in SLE patients but other studies denied this correspondence. [7,8]

The three major antigenic proteins of ribosomes, P0 (38 KD), P1 (19 KD) and P2 (17 KD) seem to share a common antigenic epitope constituted by the peptide C-terminal of the three proteins. We tested sera of 103 patients with SLE in IB on total ribosomal proteins separated electrophoretically. Among these 103 patients, 20 (19%) reacted against the three major antigenic proteins of ribosomes, P0 (38 KD), P1 (19 KD) and P2 (17 KD), but 11 of those (11%) reacted only against the sole P0, while only 3 (3%) reacted with the three proteins; 2 sera reacted with the P0 and the P1, another 2 sera reacted with the P0 and the P2, and 1 serum reacted only with either the P1 or the P2. These data mean that the shared epitope seems not to be the only antigenic determinant present in those proteins. We are now at work to verify if a different reactivity pattern corresponds to a particular clinical state.

Antinuclear Antibodies as Clues to Identify the Pathogenesis of CTD

The study of the reactivity of antinuclear antibodies may help to understand the pathogenesis of CTD. It is now well established that anti-Sm antibodies are highly specific for diagnosis of SLE. In particular anti-Sm antibodies recognize a macromolecular complex including small nuclear RNAs and a protein moiety, composed of several polypeptides (B'/B and D reacting specifically with anti-Sm antibodies). Among these polypeptides the Sm D have a molecular weight of 13 KD and is composed by 119 aminoacids. In order to understand the antigenic determinant composition of this protein, a study of epitope mapping was performed. Seven synthetic peptides covering the entire composition of the protein were synthetized (1-19, 15-34, 30-49, 45-64, 60-79, 75-94, 95-119). Then, a number of SLE sera was tested in ELISA against each one of these peptides, coated on the plates. The results showed a higher level of reactivity against the C-terminal peptide, that seems to be a major antigenic determinant of the entire protein [9].

A number of sera from patients with various CTD and infectious mononucleosis were tested against this peptide, showing that only in 30% of SLE sera and in a small number of infectious mononucleosis sera a positive reaction was demonstrated. This peptide shares an impressive homology of sequence with the peptide 35-58 of EBNA-I antigen, one of the nuclear antigens induced by Epstein Barr virus. So autoantibodies anti-SmD peptide were affinity purified, and we showed that they reacted with EBNA-I antigen. Subsequently, mice were immunized against the 35-58 EBNA-I peptide and the induced antibodies obtained displayed the same reactivity of spontaneously produced anti-SmD antibodies [10]. So molecular mimicry must be considered here as possible pathogenetic mechanism for eliciting autoantibody production.

Conclusion

During the past years the determination of nature and specificity of antinuclear antibodies has been the most important clue for the diagnosis and for the subsetting of CTD. Studies that could be able to define the intimate structure of autoantigens, might be important to disclose new insights to identify the pathogenetic mechanism of autoantibody production.

References

1. Neri R, Tavoni A, Cristofani R, et al. Antinuclear antibody profile in Italian patients with connective tissue diseases. Lupus 1992; 1:221-227.
2. Charles PJ, van Venrooij WJ, Maini RN, and the Consensus Finding Group for Autoantibodies. Clin Exp Rheumatol 1992; 10:507-511.
3. Buyon JP, Ben-Chetrit E, Karp S, et al. Acquired congenital heart block. Pattern of maternal antibody response to biochemically defined antigens of the ssa/Ro-SSB/La system in neonatal lupus. J Clin Invest 1989;84:627-634.
4. Wu DH, Tavoni A. A simple method for the biochemical purification of Ro/SS-A antigen. J Immunol Methods 1989; 121:219-34.
5. Tavoni A, Neri R, Storti S, et al. Autoantibody responses to the Ro/SS-A 60 KD and 52 KD polypeptides in systemic lupus erythematosus and primary Sjøgren syndrome. Hungarian Rhuemathology 1991; XXXII:141.
6. Scamardella F, Tavoni A, Neri R, Vitali C, Pellegrini M, Bombardieri S. Autoantibody responses to the La/SS-B polypeptides in systemic lupus erythematosus and primary Sjøgren Syndrome. Clin Rheumatol 1993; 12:40.
7. Teh LS, Bedwell AE, Isenberg DA, et al. Antibodies to protein P in systemic lupus erythematosus. Ann Rheum Dis 1992; 51:489-494.
8. Nojima Y, Minota S, Yamada A, Takaku, Aotsuka S, Yokohari R. Correlations of antibodies to ribosomal P protein with psychosis in patients with systemic lupus erythematosus. Ann Rheum Dis 1992; 51:1053-1055.
9. Sabbatini A, Dolcher MP, Marchini B, Bombardieri S, Migliorini P. Mapping of epitopes on the SmD molecule: the use of multiple antigen peptides to measure autoantibodies in systemic lupus. J Rheumatol (in press).
10. Sabbatini A, Bombardieri S, Migliorini P. Autoantibodies from patients with systemic lupus erythematosus bind a shared sequence of SmD and Epstein-Barr virus-encoded nuclear antigen EBNA I. Eur J Immunol 1993; 23:1146-1152.

TAP1 ALLELE ASSOCIATION WITH IDDM

Deborah G. Jackson and J. Donald Capra
Department of Microbiology
University of Texas Southwestern Medical Center
5323 Harry Hines Blvd.
Dallas, Texas 75235-9048
USA

Introduction

The recently identified TAP1 and TAP2 (Transporter associated with Antigen Processing) proteins share homology with such proteins as MDR (Multi Drug Resistance) protein and CFTR (Cystic Fibrosis Transmembrane conductance Regulator) and other members of the ABC superfamily of proteins which transport materials across membranes [1-4]. Although the actual function of the TAP genes is in question, TAP protein expression is required for normal levels of surface expression of class I [1]. Indirect evidence has been presented to suggest that TAP proteins function as a heterodimer to transport peptides across the ER membrane where the peptides associate with assembling class I molecules [5-9].

A variety of autoimmune disorders, including insulin dependent diabetes mellitus, are known to be associated with MHC genes [10]. It is not clear whether the destruction of the β-islet cells of the pancreas observed in IDDM is mediated through a class I or class II event. It has been observed that the majority of T lymphocytes which infiltrate the pancreas at the onset of IDDM are CD8+ [11,12]. An increase in class I expression on islet cells has also been observed. Since the TAP proteins have been implicated in class I expression and since allelic differences in the TAP genes could affect the types of peptides presented by class I molecules, an investigation of the TAP genes in relation to IDDM seems warranted.

The TAP genes are located within the 430 kb region between DQ and DP and it has not been firmly established whether they are in linkage disequilibrium with DQ or DP [1,2,13-15]. Several groups, including our own, have demonstrated an association of the DQB1*0302 allele with an increased risk for the development of insulin dependent diabetes mellitus while other alleles (i.e. DQB1*0602) were associated with protection [16-20]. In subsequent studies it was determined that there was little association with known DP alleles and susceptibility to IDDM [19].

A. G. Dalgleish et al. (eds.), The Impact of Biotechnology on Autoimmunity, 99–106.

disequilibrium between DQ and DR alleles [21]. It is not clear where within the class II region the linkage disequilibrium observed in the HLA complex breaks down. Restriction fragment length polymorphism analysis of a limited number of homozygous cell lines suggested that recombination hot spots may exist between TAP1 and DP and between TAP1 and DQ [13]. Thus, while the apparent relative risk for IDDM appears to peak at the DQ locus, it is possible that a more significant association might be observed at the TAP locus which is located more centromeric on the 6th chromosome.

Methods

PATIENTS AND NORMALS

One hundred and forty nine unrelated Caucasian patients with IDDM as defined by the criteria of the National Diabetes Data Group were studied. The mean age of the patients at the onset of their diabetes was 9.3 years and the mean duration of the disease was 8.2 years. One hundred and forty random unrelated Caucasian individuals served as normal controls. Each described himself or herself as non-diabetic, but blood glucose was not measured in these subjects. Other details of the patient and control population are as described previously [18].

SINGLE STRANDED CONFORMATIONAL POLYMORPHISM [SSCP] ANALYSIS

SSCP oligonucleotides (16 μg) were labeled in a 50 μL reaction with 300 μCi ^{32}P(specific activity 3000 Ci/mMole) using 16U T4 polynucleotide kinase (Promega) for 1 hour at 37^{0}C. Labeled oligonucleotides were precipitated and diluted to 0.16 μg/μL. PCR reaction conditions for a 100 μL reaction were: 0.32 μg of primers, 1 μg DNA,1X Taq buffer, 200 μM dNTPs and 1U of Taq polymerase. PCR protocols included a 5 minute step of 94^{0}C, followed by 30 cycles of 94^{0}C for 1 minute, annealing temperature for 2 minutes and 72^{0}C for 3 minutes.

SSCP GELS

SSCP gels were prepared using 2X MDE acrylamide. MDE is a reagent manufactured by J.T. Baker in a 2X concentration. A final concentration of either 0.70X (2110-2230/intron), 0.5X or 0.45X (1501-1840bp, 1879-2201 and 1058-1260/intron) MDE in 0.6X TBE was used for identification of alleles. Gels were run at 10 watts constant power for 9-12 hrs. SSCP-PCR reactions were diluted 1:10 in 95% formamide/10mM NaOH, 0.025% bromphenol blue and 0.025% xylene cynol. Samples were heated to 95^{0}C for 5 minutes then placed on ice for 1-2 minutes prior to loading. Typically 3 μl of sample was loaded in each well. Three sets of reactions were used to determine the allele type. Gels were dried for 1 hour under vacuum

Samples were heated to 95°C for 5 minutes then placed on ice for 1-2 minutes prior to loading. Typically 3 μl of sample was loaded in each well. Three sets of reactions were used to determine the allele type. Gels were dried for 1 hour under vacuum and placed under film for 4 to 24 hours.

STATISTICAL ANALYSIS

Gene frequencies were calculated from the frequencies at which the alleles were observed using the formula $g=\sqrt{1-f}$, where f is the observed allele frequency and g is the gene frequency. Since a large number of unrelated individuals were tested, no distinction was made for homozygosity for a particular allele. The significance of the differences in allele frequencies was determined by Chi-square analysis. Relative risk was calculated by the method of Woolf with the use of the formula (axd)/(bxc). When one element of the equation was zero, the relative risk was calculated with the formula of Haldane:

$$[(2a+1)(2d+1)]/[(2b+1)(2c+1)].$$

Results

Previous studies had suggested that polymorphisms of the TAP genes were localized to the C-terminal half of the molecule [4,22,23]. Genomic sequencing in our laboratory confirmed that the 5' half of the TAP1 gene was non-polymorphic. Thus we concentrated most of our investigation on the 3' half of the gene. TAP1 cDNA clones were generated from the lymphocytes of 5 diabetics and 11 nondiabetics. Sequencing 1250-1300 bp of 5-10 clones from these 16 individuals resulted in the identification of four unique sequences. The resulting amino acid substitutions are summarized in Table 1. TAP1*0101[1] was observed in all but one of the individuals studied by sequencing. One diabetic was homozygous for TAP1*0301 which contains a val-333 resulting from an A to G substitution at nucleotide 1207. Two nondiabetics were found to be heterozygous for TAP1*0301 . TAP1*0201 was identified in one individual. In addition to the val-333 amino acid substitution observed in TAP1*0301, TAP1*0201 contains an A to G base change at nucleotide 2120 which results in the amino acid corresponding to position 637 changing from an asp to a gly. Sequence analysis of clones generated from ten members of two families resulted in the identification of TAP1*0401, which contains the val-333, the gly-637 and two additional unique changes, a G to T base change resulting in a val to leu substitution at amino acid 458 and a G to A base change which results in an arg to gln substitution at position 648. At least 50% of the clones from 5 members of the two families possessed these four amino acid changes. By sequencing the

---

[1]Provisionally we have called the alleles TAP1*0101, *0201, *0301, *0401, *0501. Request for approval of this nomenclature has been submitted to the World Health Organization (WHO) nomenclature committee.

generated from both cDNA and genomic DNA confirmed that the results from the clones were not due to Taq polymerase or reverse transcriptase errors. An additional (sixth) allele referred to a TAP1 1C by Powis et al. [23] was not observed during this investigation.

Table 1. Amino Acid Substitutions in TAP-1 Alleles

AMINO ACID	333		458		637		648	
NUCLEOTIDE	1207		1582		2120		2153	
	ATC	GTC	GTG	TTG	GAC	GGC	CGA	CAA
TAP-1*0101	ILE		VAL		ASP		ARG	
TAP-1*0201		VAL	VAL			GLY	ARG	
TAP-1*0301		VAL	VAL		ASP		ARG	
TAP-1*0401		VAL		LEU		GLY		GLN

Nucleotide numbering is based on the cDNA sequence of Trowsdale et al. [1], amino acid numbering is based on Colonna et al. [22]

Using this sequence information, SSCP analysis was used to screen cDNA of other patients with IDDM and normals. Oligonucleotides primers derived from cDNA sequence were generated to produce 300 to 400 bp fragments covering the entire human TAP1 cDNA. Analysis of fragments generated from the 5' half of the cDNA indicated that there were no variations in this region. Sequence analysis of the SSCP fragment (900-1240 bp) verified that the observed SSCP pattern was always due to the ile to val change at amino acid 333. SSCP analysis of two specific families documented that val-333, leu-458, gly-637 and gln-648 always segregated together. No other variants were apparent from SSCP analysis of cDNA fragments derived from amplification of other regions of TAP1.

Ninety-seven patients with IDDM that had been previously typed at DRβ1, DQα1, DQβ1 and DPβ1 and an additional 52 diabetic DNA samples typed at DRβ1, DQβ1 and DQα1 but not DPβ1 were screened for TAP1 using genomic SSCP. Genomic DNA samples from 140 normal Caucasians were also screened for TAP1 alleles by SSCP. The TAP1 allele was identified by analysis of three genomic SSCP fragments. Analysis of the third fragment (2110-2230) identified a fifth TAP1 allele which was designated TAP*0501. TAP1*0501 contains val-333, gly-637 and an addition change between 2110 and 2230 bp. Gene frequencies for the five TAP1 alleles are summarized in Table 2. The gene frequency of TAP1*0301 is approximately 7 fold increased in patients with IDDM and is therefore associated a relative risk of 8.3. The gene frequency of TAP1*0201 is slightly increased in the diabetic population and is associated with a risk of 2.1.

Table 2. TAP1 Gene Frequencies

	Normal (N=140)	IDDM (N=149)	Relative Risk
TAP*0101	138 (88.0)	137 (72.0)	0.2*
TAP*0201	17 (6.27)	33 (11.8)	2.1
TAP*0301	3 (1.07)	23 (8.04)	8.3**
TAP*0401	11 (4.00)	4 (1.35)	0.3
TAP*0501	10 (3.63)	4 (1.35)	0.4
BLANK	- (-2.9)	- (5.46)	

Alleles typed in normals -179
Alleles typed in IDDM -201
g= gene frequency, f= observed allele frequency, ($g-1-\sqrt{1-f}$)
Homozygotes were not considered in calculation (see Methods)
* $X^2=6.8729$, corrected $p<0.05$, ** $X^2=15.5795$, corrected $p<0.0005$

Homozygotes of TAP1*0301 and TAP1*0201 were observed in the diabetic population but not in the normal population. The TAP1*0301/TAP1*0301 homozygous genotype was observed in 3 diabetics (2.0%) and is associated with a relative risk of 6.7 (Table 3). The TAP1*0301/TAP1*0101 heterozygous genotype is associated with a significant relative risk (5.1) but represents only 15.4% of the diabetics. An excess in observed TAP*0101 homozygotes in the diabetic population over the expected number based on calculated gene frequencies was also observed. The normal population was found to be in Hardy Weinberg equilibrium with respect to TAP1.

Discussion

The IDDM association observed with TAP1 parallels the results obtained for the DP locus [20]. While TAP1*0301 is associated with a high relative risk (8.3) the etiologic fraction of individuals with TAP1*0301 is relatively small (15.4%). The other TAP1 alleles are associated with either moderate relative risk or are protective. TAP1*0301 does not group any specific diabetics from the perspective of their DQα or β type or their DPβ type. Specifically, had our analysis identified for example that 5% of Caucasians who are neither DR3 or DR4 or the group who are not DQB1*0302 or DQB1*0201 this might have been of considerable interest. On the other hand analysis of TAP1*0301 individuals shows that their distribution of DQβ alleles is no different that the distribution of the more common TAP1*0101 individuals (data not shown). These results suggest that while there may be an

Table 3. Frequencies of TAP1 Alleles in normal subjects and patients with IDDM

	Normals (N=140)	IDDM (N-149)	Relative Risk
0101/0101	100	92	0.6
0101/0301	3	15	5.1*
0101/0201	16	25	1.5
0101/0401	10	2	0.2**
0101/0501	9	3	0.1
0301/0301	0	3	6.7
0201/0301	0	4	-
0301/0401	0	1	-
0301/0501	0	0	-
0201/0201	0	2	-
0201/0401	0	1	-
0201/0501	1	1	-
0401/0401	1	0	-
0401/0501	0	0	-
0501/0501	0	0	-

- too few normals/patients to calculate
*X^2=7.7605, p=0.0053; **X^2=6.1023, p=.0135

association of IDDM with TAP1*0301, it is most likely the result of a passenger effect related to a gene that has yet to be identified located between TAP1 and DQβ1 or that DQβ1 itself is responsible for the HLA associated susceptibility to IDDM. It has been shown through studies of deletion mutant cell lines that at least one unknown gene in the vicinity of the TAP genes is needed for the intracellular production of class II-peptide complexes [24]. Such a gene or genes could dramatically impact expression of self reactive peptides presented by class II and could be tightly linked to any disease association related to class II expression.

References

1. Trowsdale J, Hanson I, Mockridge I, Beck S, Townsend A, Kelly A. Sequences encoded in the class II region of the MHC related to the "ABC" superfamily of transporters. Nature (London) 1990;348:741-743.
2. Spies T, Bresnahan M, Bahram S, Arnold D, Blanck G, Mellins E, Pious D, DeMars R. A gene in the human major histocompatibility complex class II region controlling the class I antigen presentation pathway. Nature (London) 1990;348:744-747.
3. Bahram S, Arnold D, Bresnahan M, Strominger J, Spies T. Two putative subunits of a peptide pump encoded in the human major histocompatibility complex class II region. Proc Natl Acad Sci 1991;88:10094-10098.
4. Powis SH, Mockridge I, Kelly A, Kerr L, Glynne R, Gileadi U, Beck S, Trowsdale J. Polymorphisms in a second ABC transporter gene located within the class II region of the human major histocompatibility complex. Proc Natl Acad Sci 1992;89:1463-1467.
5. Spies T, Cerlundolo V, Colonna M, Cresswell P, Townsend A, DeMars R. Presentation of viral antigen by MHC class I molecules is dependent on a putative peptide transporter heterodimer. Nature 1992;355:644-646.
6. Monaco J. Genes in the MHC that may affect antigen processing. Current Opinion in Immunology 1992;4:70-73.
7. Monaco J. A molecular model of MHC class I restricted antigen processing. Immunology Today 1992;13:173-179.
8. Spies T, DeMars R. Restored expression of major histocompatibility class I molecules by gene transfer of a putative peptide transporter. Nature (London) 1991;351:323-324.
9. Kleijmeer MJ, Kelly A, Geuze HJ, Slot JW, Townsend A, Trowsdale J. Location of MHC-encoded transporters in the endoplasmic reticulum and cis-golgi. Nature (London) 1992;357:342-344.
10. Nepom G, Erlich H. MHC class II and autoimmunity. Annual Rev Immunol 1991;9:493-525.
11. Bottazzo GF, Dean BM, McNally JM, MacKay EH, Swift PGF, Gamble DR. In situ characterization of autoimmune phenomena and expression of HLA molecules in the pancreas in diabetic insulitis. N Engl J Med 1985;313:353-360.
12. Hanninen A, Jalkanen S, Salmi M, Toikkanen S, Nikolakaros G, Simell O. Macrophages, T cell receptor usage, and endothelial cell activation in the pancreas at the onset of insulin dependent diabetes mellitus. J Clin Invest

1992;90:1901-1910.

14. van Endert PM, Lopez MT, Patel SD, Monaco JJ, McDevitt HO. Genomic polymorphism, recombination, and linkage disequilibrium in human major histocompatibility complex encoded antigen-processing genes. Proc Natl Acad Sci 1992;89:11594-11597.
15. Trowsdale J, Ragoussis J, Campbell RD. Map of the human MHC. Immunology Today 1991;12:443-444.
16. Todd JA, Bell JI, McDevitt HO. HLA-DQβ contributes to susceptibility to and resistance to insulin dependent diabetes mellitus. Nature (London) 1987;329:599-604.
17. Sterkers G, Zeliszewski D, Chaussee AM, Deschamps I, Font MP, Freidel C, Hors J, Betuel H, Dausset J, Levy JP. HLA-DQ rather than HLA-DR region might be involved in dominant nonsusceptibility to diabetes. Proc Natl Acad Sci 1988;85:6473-6477.
18. Baisch JM, Weeks T, Giles R, Hoover M, Stastny P, Capra JD. Analysis of HLA-DQ genotypes and susceptibility in insulin dependent diabetes mellitus. New Eng J of Med 1990;322:1836-1841.
19. Dorman JS, LaPorte RE, Stone RA, Trucco M. Worldwide differences in the incidence of type I diabetes are associated with amino acid variation at position 57 of the HLA-DQβ chain. Proc Natl Acad Sci 1990;87:7370-7374.
20. Baisch JM, Capra JD. Analysis of HLA genotypes and susceptibility to Insulin-dependent diabetes mellitus: association Maps telomeric to HLA-DP. Scand J Immunology 1990;36:331-340.
21. Ronningen KS, Spurkland A, Iwe T, Vartdal F, Thorsby E. Distribution of HLA-DRB1, -DQA1 and -DQB1 alleles and DQA1-DQB1 genotypes among Norwegian patients with insulin-dependent diabetes mellitus. Tissue Antigens 1991;37:105-111.
22. Colonna M, Bresnahan M, Bahram S, Strominger J, Spies T. Allelic variants of the human putative peptide transporter involved in antigen processing. Proc Natl Acad Sci USA 1992;89:3932-3936.
23. Powis S, Tonks S, Mockridge I, Kelly AP, Bodmer JG, Trowsdale J. Alleles and haplotypes of the MHC-encoded ABC transporters TAP1 and TAP2. Immunogenetics 1993;37:373-380.
24. Ceman S, Rudersdorf R, Long EO, Demars R. MHC class II deletion mutant expresses normal levels of transgene encoded class II molecules that have abnormal conformation and impaired antigen presentation ability. J Immunol 1992;149:754-761.

INDUCTION OF PERIPHERAL TOLERANCE IN PRIMED MICE

Luciano Adorini and Jean-Charles Guéry
Roche Milano Ricerche
Via Olgettina, 58
I-20132 Milano
ITALY

Summary

Proteins or synthetic peptides administered in soluble form can induce peripheral T cell tolerance. Hen egg-white lysozyme (HEL) delivered continuously over a 10-day period by an osmotic minipump implanted subcutaneously, efficiently and specifically prevents the subsequent priming of T cells by injection of HEL emulsified in adjuvant. Similar results can be obtained by administration of soluble HEL peptides. T cells specific for dominant or subdominant HEL epitopes can be made tolerant equally well. Tolerance induction prevents effectively the helper function of Th1 but not Th2-type cells, as documented by *in vivo* inhibition of anti-HEL antibody responses of IgG2a, but not IgG1 isotype. Addition of IL-2 to cultures from tolerant mice restores T cell responsiveness, indicating anergy rather than deletion of T cells as the mechanism of tolerance induction. Mixing experiments fail to demonstrate induction of suppressor T cells in tolerant mice. Tolerance is long-lasting, and it is still present 60 days after administration of antigen in tolerogenic form. Interestingly, tolerance, rather than a secondary response, is induced by continuous antigen administration to antigen-primed mice, suggesting the possible applicability of this form of tolerance induction to selective immunosuppression of established autoimmune diseases.

Introduction

The essence of the immune system is to distinguish self from nonself and this discrimination is primarily carried out by $CD4^+$ T cells. However complex as to the mechanisms utilized, self-nonself discrimination is teleologically simple: it positively selects T cells capable of recognizing nonself while eliminating, physically or functionally, T cells capable of responding to self antigens. Tolerance to self is explained by three major mechanisms. First, T cells expressing receptors specific for self antigens can be physically eliminated during thymic development, leading to

A. G. Dalgleish et al. (eds.), The Impact of Biotechnology on Autoimmunity, 107–114.

clonal deletion [18,19]. Although this is a primary mechanism of self tolerance, it does not eliminate T cells specific for antigens not expressed on thymic antigen-presenting cells (APC). Second, self reactive T cells that have been exported in the periphery can be functionally inactivated upon antigen recognition in the absence of an appropriate costimulatory signal, a process denominated clonal anergy [17]. Third, peripheral self-reactive T cells could be persistently suppressed by other T cells [21]. In reality, self-nonself discrimination is truly redundant, and both deletional and non-deletional mechanisms operate in central and in peripheral tolerance [16].

Failure of self-tolerance may result in autoimmune diseases. Therefore, the most effective and selective form of immunointervention in autoimmune diseases would be to establish or re-establish tolerance to self antigens [3]. This requires knowledge of the autoantigen recognized by pathogenic T cells, at present poorly defined in most autoimmune diseases. Progress is however expected in the identification of reliable autoantigen candidates, and this could allow induction of peripheral tolerance to the autoantigen.

Antigen-specific unresponsiveness has been induced *in vivo* by intravenous administration of soluble antigen [11], antigen-pulsed, chemically modified syngeneic cells [17], hapten-coupled syngeneic cells [4], allogeneic or Mls-disparate cells [25]. In general, tolerance induction by intravenous antigen administration takes place in the periphery, since it can be established in athymic mice, and it appears to be mediated more frequently by clonal anergy [5,25] rather than by clonal deletion [27]. However, in all these models tolerance has only been induced in unprimed individuals, since bolus antigen administration would induce a secondary response, rather than tolerance, in an antigen-primed immune system. Induction of specific unresponsiveness in antigen-primed mice could offer the possibility to treat established autoimmune diseases by inducing tolerance in activated T cells specific for self antigens.

Induction of T Cell Tolerance by Administration of Soluble Proteins

Administration of soluble proteins readily induces T cell tolerance. A representative experiment is shown in Figure 1. C3H mice were implanted with a mini-osmotic pump delivering continuously for approximately 10 days about 0.5 nmoles/hr of HEL or ribonuclease (RNase). Twelve days after pump implantation mice were primed into the hind footpads with HEL or RNase emulsified in CFA, and 8 days later antigen-specific T cell proliferative responses were measured in popliteal lymph node cells from individual mice. The results demonstrate that delivery of soluble HEL induces a profound inhibition of HEL-specific T cell proliferation, whereas the response to PPD is unaffected. As a further specificity control, soluble HEL did not interfere with the T cell proliferative reponse to RNase, whereas soluble RNase was very effective in inhibiting selectively the anti-RNase T cell response. Again, the response to PPD was not affected.

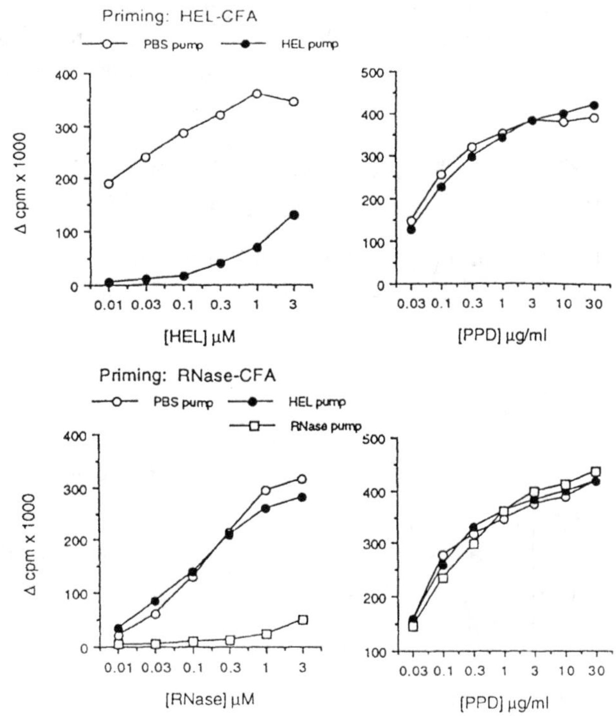

Figure 1. Tolerance induction by continuous administration of soluble proteins. Results refer to mean T cell proliferative responses of lymph node cells from 3 mice/group.

We next determined the inhibition of T cell responses to dominant and subdominant HEL epitopes by administration of soluble HEL to C3H mice. The set of overlapping HEL peptides used defines in mice of H-2^k haplotype three dominant epitopes, included in HEL sequences 1-18, 46-61, and 107-129. The T cell response to the former epitope is restricted by I-E^k molecules, whereas the latter two are I-A^k-restricted [2]. A T cell response is also observed to the subdominant HEL epitope 25-43. Administration of soluble HEL inhibits very effectively T cell responses to both dominant and subdominant HEL epitopes. This result is quite interesting, since a point of debate is the dominance of autoantigenic peptides presented by class II molecules associated with autoimmune disease susceptibility. Acccording to one theory, autoreactive pathogenic T cell clones could recognise cryptic determinants, and for this reason have escaped thymic deletion during negative selection [13]. The opposite view considers autoantigenic peptides as dominant, and therefore highly antigenic. When events such as molecular mimicry [24] or regulatory imbalances [23] lead to breakdown of tolerance, their antigenic potential can be expressed. Our results, demonstrating effective inhibition of T cell responses induced by dominant and subdominant antigenic epitopes, would indicate that in either case induction of peripheral T cell tolerance may be capable of interfering with the activation of autoreactive T cells.

Inhibition of *In Vivo* Anti-HEL Antibody Responses by Induction of T Cell Tolerance to HEL

Tolerance induction by administration of soluble proteins inhibits not only T cell responses, but also T cell-dependent antibody responses *in vivo* (Figure 2). Two

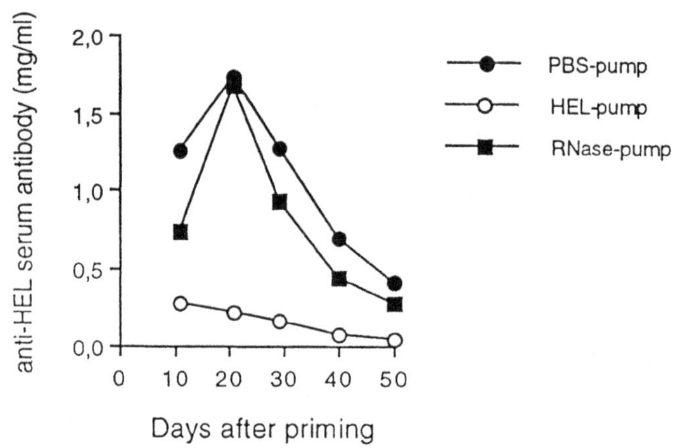

Figure 2. Inhibition of *in vivo* anti-HEL antibody responses by induction of T cell tolerance to HEL. C3H mice were implanted with a mini-osmotic pump delivering continuously for approximately 10 days about 0.5 nmoles/hr of HEL or ribonuclease (RNase). Eighteen days after pump implantation mice were primed i.p. with 35 pmoles of HEL emulsified in CFA. Sera were collected at the indicated days after priming and total anti-HEL antibodies quantitated by ELISA as described [15].

types of helper T cells, Th1 and Th2, have been described in the mouse, based on their different lymphokine production pattern [22]. Th1 cells, which produce IL-2 and IFN-g, induce preferentially synthesis of antibodies expressing the IgG2a isotype. Conversely, Th2 cells, producing IL-4, IL-5 and IL-10, mainly elicit antibody responses of IgG1 and IgE isotype and cause B cell proliferation and differentiation

[1,7,12,26].

Our protocol of tolerance induction inhibits very efficiently (about 100-fold decrease) IgG2a/IgG2b isotypes, whereas anti-HEL antibodies of IgG1 isotype show only a 5-fold reduction (Table 1). These results indicate that mostly Th1 rather than Th2-type helper cells are inhibited by tolerance induction, consistent with the observation that antigen-specific unresponsiveness induced by pretreatment with aqueous antigen selectively tolerizes Th1-like cells [6,9]. Since Th1-type cells are the pathogenic cells in autoimmune diseases, tolerance induction of this T cell subset may interfere with the course of disease. Conversely, human CD4$^+$ T cell clones specific for allergenic and helmintic antigens exhibit Th2-like lymphokine production profile [10], and in these cases tolerance induction may not affect pathogenic T cells.

Table 1. Preferential inhibition of Th1-dependent anti-HEL isotypes in C3H mice made tolerant of HEL.

		Serum antibody concentration (U/ml ± SEM)			
Pump	Priming	Total IgG	IgG1	IgG2a	IgG2b
no antigen	HEL	1,764 ± 215	1,340 ± 138	54.3 ± 8.2	28.8 ± 7.9
HEL	HEL	183 ± 9.4 (9x)	241 ± 30.6 (5x)	0.38 ± 0.08 (144x)	0.35 ± 0.19 (82x)

Sera from an experiment similar to that described in Figure 1 were tested for anti-HEL antibody isotypes by quantitative ELISA [15]. Results are expressed as serum antibody concentration (U/ml) and SEM of 3-4 mice/group.

Tolerance Induction in Primed Mice

Induction of peripheral tolerance by administration of autoantigens has been demonstrated in autoimmune models. Experimental allergic encephalomyelitis (EAE) can be ameliorated by inducing T cell anergy to synthetic peptides corresponding to the immunodominant epitopes of myelin basic protein (MBP) [14]. In this case, tolerance induction by injection of MBP peptides emulsified in IFA at EAE onset, 10 days after administration of MBP, blocked the progression and decreased the severity of EAE. These results suggest that selective immunotherapy by administration of autoantigenic peptides can be effective in acute experimental autoimmune diseases, even after the autoimmune response has been initiated. However, in chronic diseases amplification and broadening of the autoimmune response to other epitopes of the same autoantigen appears to occur [20], suggesting that induction of immunosuppression by immunodominant peptides only [8] may not be effective. This problem could be circumvented by tolerance induction to the entire autoantigenic protein.

Therefore, we wished to determine whether continuous administration of soluble protein antigen could induce tolerance rather than a secondary response in primed mice. Preliminary results demonstrate that continuous administration of soluble HEL to HEL-primed mice is still capable of inhibiting HEL-specific T cell proliferative responses, indicating the possibility of inducing peripheral tolerance by administration of intact proteins to primed mice.

References

1. Abbas AK, Williams ME, Burstein HJ, Chang T-L, Bossu P, Lichtman AH. Activation and functions of CD4+ T-cell subsets. Immunol Rev 1991;123: 5-22.
2. Adorini L, Appella E, Doria G, Nagy ZA. Mechanisms influencing the immunodominance of T cell determinants. J Exp Med 1988;168:2091-2104.
3. Adorini L, Guéry J-C, Rodriguez-Tarduchy G, Trembleau S. Selective immunosuppression. J Exp Med 1993;177:1461-1468.
4. Battisto JR, Bloom BR. Dual immunological unresponsiveness induced by cell membrane coupled hapten or antigen. Nature 1966;212:156.
5. Burkly LC, Lo D, Kanagawa R, Brinster L, Flavell RA. T cell tolerance by clonal anergy in transgenic mice with nonlymphoid expression of MHC class II I-E. Nature 1989;342:564.
6. Burstein HJ, Shea CM, Abbas AK. Aqueous antigens induce in vivo tolerance selectively in IL-2- and IFN-g-producing (Th1) cells. J Immunol 1992;148:3687-3691.
7. Burstein HJ, Tepper RI, Leder P, Abbas AK. Humoral immune functions in IL-4 transgenic mice. J Immunol 1991;147:2950-2956.
8. Clayton JP, Gammon GM, Ando DG, Kono DH, Hood L, Sercarz EE. Peptide-specific prevention of experimental allergic encephalomyelitis. Neonatal tolerance induced to the dominant T cell determinant of myelin basic protein. J Exp Med 1989;169:1681-1691.
9. De Wit D, Van Mechelen M, Ryelandt M, et al. The injection of deaggregated gamma globulins in adult mice induces antigen-specific unresponsiveness of T helper type 1 but not type 2 lymphocytes. J Exp Med 1992;175:9-14.
10. Del Prete G, De Carli M, Mastromauro C, et al. Purified protein derivative of mycobacterium tuberculosis and escretory-secretory antigen(s) of toxocara canis expand in vitro human T cells with stable and opposite (type 1 T helper or type 2 T helper) profile of cytokine production. J Clin Invest 1991;88:346-350.

11. Dixon FJ, Maurer PH. Immunological unresponsiveness induced by protein antigens. J Exp Med 1955;101:245.
12. Finkelman FD, Holmes J, Katona IM, et al. Lymphokine control of in vivo immunoglobulin isotype selection. Annu Rev Immunol 1990;8:303-333.
13. Gammon G, Sercarz E. How some T cells escape tolerance induction. Nature 1989;342:183-185.
14. Gaur A, Wiers B, Liu A, Rothbard J, Fathman CG. Amelioration of autoimmune encephalomyelitis by myelin basic protein synthetic peptide-induced anergy. Science 1992;258:1491-1494.
15. Guéry J-C, Neagu M, Rodriguez-Tarduchy G, Adorini L. Selective immunosuppression by administration of major histocompatibility complex class II-binding peptides. II. Preventive inhibition of primary and secondary antibody responses. J Exp Med 1993;177:1461-1468.
16. Haemmerling JG, Schoenrich G, Momburg F, et al. Non-deletional mechanisms of peripheral and central tolerance: studies with transgenic mice with tissue-specific expression of a foreign MHC class I antigen. Immunol Reviews 1991;122:47-67.
17. Jenkins MK, Schwartz RH. Antigen presentation by chemically modified splenocytes induces antigen-specific T cell unresponsiveness in vitro and in vivo. J Exp Med 1987;165:302.
18. Kappler JW, Roehm N, Marrack P. T cell tolerance by clonal elimination in the thymus. Cell 1987;49:273.
19. Kisielow P, Bluethmann H, Staerz UD, Steinmetz M, von Boehmer H. Tolerance in T cell receptor transgenic mice involves deletion of nonmature CD4+8+ thymocytes. Nature 1988;333:742.
20. Lehmann PV, Forsthuber T, Miller A, Sercarz EE. Spreading of T-cell autoimmunity to cryptic determinants of an autoantigen. Nature 1992;358:155-157.
21. Lo D, Burkly LC, Flavell RA, Palmiter RD, Brinster RL. Tolerance in transgenic mice expressing class II major histocompatibility complex on pancreatic acinar cells. J Exp Med 1989;170:87.
22. Mosmann TR, Coffman RL. Th1 and Th2 cells: Different patterns of lymphokine secretion lead to different functional properties. Ann Rev Immunol 1989;7:145-173.
23. Nepom GT. MHC class II molecules and autoimmunity. Annu Rev Immunol 1991;9:493-520.
24. Oldstone MB. Molecular mimicry as a mechanism for the cause and a probe uncovering etiologic agent(s) of autoimmune diseases. Curr Top Microbiol Immunol 1989;145:127-139.
25. Rammensee H-G, Kroschewski R, Frangoulis B. Clonal anergy induced in mature $Vb6^+$ T lymphocytes on immunizing $Mls-1^b$ mice with $Mls-1^a$ expressing cells. Nature 1989;339:541-544.
26. Snapper CM, Paul WE. Interferon-g and B cell stimulatory factor-1

reciprocally regulate Ig isotype production. Science (Wash. DC) 1987;236:944-947.

27. Webb S, Morris C, Sprent J. Extrathymic tolerance of mature T cells: clonal elimination as a consequence of immunity. Cell 1990;63:1249.

EPITOPE SCREENING LIBRARIES: A NEW STRATEGY TO IDENTIFY MHC-SPECIFIC ANTAGONISTS

Francesco Sinigaglia and Juergen Hammer
Roche Milano Ricerche
Via Olgettina, 58
I-20132 Milan
ITALY

Summary

The major histocompatibility complex class II molecules (MHC) are highly polymorphic membrane glycoproteins that bind peptide fragments of proteins and display them for recognition by $CD4^+$ T cells. To understand the role of human MHC class II polymorphic residues that are associated with susceptibility to autoimmune diseases, we have isolated M13 phage from a large M13 peptide display library by selection with the rheumatoid arthritis (RA) - associated DRB1*0101 and DRB1*0401 and the RA-non associated DRB1*1101 molecules. Sequence analysis of the peptide encoding region of DR-bound phage led to the identification of position specific anchor residues, defining motifs for peptide binding to DR molecules. The three DR motifs share two anchor residues at relative position 1 and 4, while allele specific anchor residues have been identified at position 6.

Introduction

Autoimmune diseases result from the activation of self-reactive T cells by autoantigens. A striking characteristic of autoimmune diseases is the increased frequency of certain HLA class II alleles in affected individuals. Moreover, as demonstrated for example in rheumatoid arthritis and insulin-dependent diabetes mellitus, alleles positively associated with autoimmune diseases share unique amino acid residues in the hypervariable regions involved in peptide binding [1]. Although the pathogenesis of autoimmune diseases is not known, one possibility is that disease-associated HLA class II molecules have the capacity to bind and present autoantigenic peptides to T cells [2]. These peptides activate autoreactive CD4 (helper) T cells, which in turn induce a cascade of chronic inflammation.

A. G. Dalgleish et al. (eds.), The Impact of Biotechnology on Autoimmunity, 115–121.

Blocking the antigen-presenting activity of disease-associated HLA class II molecules could represent a suitable target to prevent and possibly to treat autoimmune diseases [3]. Because the autoantigens involved are largely unknown, the antigen presenting class II molecules lend themselves as the only targets available for a selective immunosuppressive therapeutic approach. Studies in the past five years have elucidated the three dimensional structure of class I [4-6], and in part, also that of class II MHC molecules [7]. It has also been demonstrated that class II MHC antagonism leads to suppression of murine T cell responses *in vivo* [8] and to prevention of disease in two different autoimmune disease models [9,10]. These results render MHC molecules feasible as pharmacological targets.

The Phage Library Approach

To develop MHC-specific antagonists it is necessary to learn the rules governing peptide-MHC interactions. In the past two years we have developed a new approach to define the characteristics of peptides binding to class II molecules by the screening of a large highly diverse peptide library expressed on the surface of filamentous phage [11].

Epitope libraries comprise tens to hundreds of millions of short, variable amino acid sequences that are displayed on the surface of a bacteriophage virion [12-14].

To construct the peptide display library we placed oligonucleotides encoding for random peptides of nine amino acid residues in length on the NH_2 terminus of the M13 phage protein III. The protein III is a minor coat protein of the M13 phage and the region containing the peptide inserts is known to be exposed on the phage surface [15]. The peptide sequences were flanked by two glycine spacers to maximize the flexibility of the peptide insert. The two RA-associated DR alleles: DRB1*0101 and DRB1*0401 and the non RA-associated DRB1*1101 allele were affinity purified from human HLA-DR homozygous EBV-B cell lines, biotinylated and used to select phage from an M13 peptide display library (Figure 1). The library, consisting of about 20 million independent phage recombinants, was screened by biopanning by reacting biotinylated DR molecules with the library, then purifying the phage that display ligand epitopes by immobilizing complexes of phage bound to biotinylated DR on streptoavidin, washing away free phage, and eluting the bound phage with acid. The eluted phage are amplified by growth in bacterial cells, then used in subsequent rounds of biopanning to further enrich for binding phage. Thus, binding phage are selectively enriched and amplified by repeated rounds of biopanning followed by infection and propagation of enriched phage. Individual phage clone were then isolated and the sequences encoding the displayed peptides were determined.

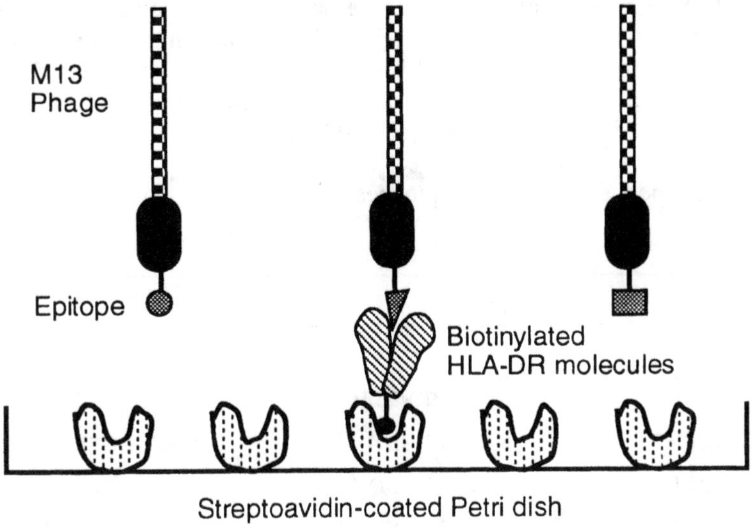

Figure 1. Schematic representation of the biopanning method to affinity purify phage that bind biotinylated DR molecules. Three phage clones are shown, each bearing a different peptide sequence fused to pIII. Phage from a library are reacted with biotinylated legate, then immobilized on a streptoavidin-coated plate and extensively washed to remove unbound or weakly binding phage. The bound phage are eluted from the plate in acid. The phage are then used to infect cells and propagated. The resulting amplified phage can be used in further rounds of biopanning, or cloned, sequenced in the epitope region and tested for binding to ligate.

Identification of Anchor Positions Involved in Peptide Binding to RA-Associated DR Alleles

Sequence analysis of the peptide encoding region of DR-bound phage led to the identification of position specific anchor residues, defining motifs for peptide binding to DR molecules (Table 1).

The three DR motifs shared two anchor residues at relative position 1 and 4, while allele specific anchor residues have been identified at position 6 [16] (Table 1).

Each of the aligned sequence pools showed a striking enrichment (>80%) of aromatic residues near the NH_2-terminus of the peptide inserts. A second general anchor residue was found at relative position 4 where the aligned peptide pools showed an enrichment of hydrophobic residues. Further constraints were found at position 6. In the DRB1*0401 selected peptides we found mostly amino acids having hydroxyl groups (Thr and Ser). The positively charged amino acids Arg and Lys were enriched in the DRB1*1101 selected pool and small amino acids (Ala and Gly) were found for the DRB1*0101 binding peptides. Such results show that the anchor residues at position 6 play an important role in determining the allele specificity of peptide binding for the HLA-DR molecules examined.

Table 1. Aligned amino acid sequences of the random peptide region of DR selected phage.

DRB1*0101	DRB1*0401	DRB1*1101
YRQMSAPTL	GWGMMRTLR	MWQELVRIQ
YKPLWGQMT	KAWSRMGSM	TWGGLGRLL
YLRMLQILK	NWRGVLSQM	LWAQVSRAI
MYGLMLAQR	WNYISSVQS	WRWMSRVSS
YKLLRATQM	WRIAQTNHA	RWKLMQKQL
YRGMSAFRA	WSTARTLWQ	TWRTLWRQV
YKPMLASVG	YHAHRTLLQ	YTQMMRSFH
NWRALRGAL	YRHAVGQLG	GWQLTAKRT
YQALRAYWQ	AWAHMTTLR	WVSLGRLAA
RVFQNLRAS	WQNMVTTLQ	MWEMLWRPR

Table 1 lists the aligned sequences of the 9mer peptide encoding region of 10 independent phage selected by each of the three DR alleles.

The phage were selected from the M13 peptide display library by DRB1*0101, DRB1*0401, and DRB1*1101 molecules and isolated after three rounds of screening [16]. Ten representative sequences for each DR allele are shown. Anchor residues are indicated in bold types.

Conserved and Allele-specific Anchors

Comparing the motifs identified for the three alleles, it is apparent that while the anchors at position 1 and 4 are practically invariant, the anchors at position 6 differ among the three DR alleles. This suggests that residues at this position confer allelic specificity to the binding. If correct, this hypothesis would predict that by exchanging only residues at position 6 we should be able to modify the allele specificity of peptide binding. To test this hypothesis, we designed and synthesized a series of peptides based on the DRB1*0101 binding motif where position 6 was replaced by anchor residues displayed by peptides binding to other DR alleles. As seen in Figure 2a the designer peptide with threonine at position 6 bound to DRB1*0401, while weak or no binding was detected for peptides carrying alanine or arginine at this position (Figure 2a). Conversely, while the arginine containing peptide bound to DRB1*1101, no binding could be detected with the other two variants (Figure 2b).

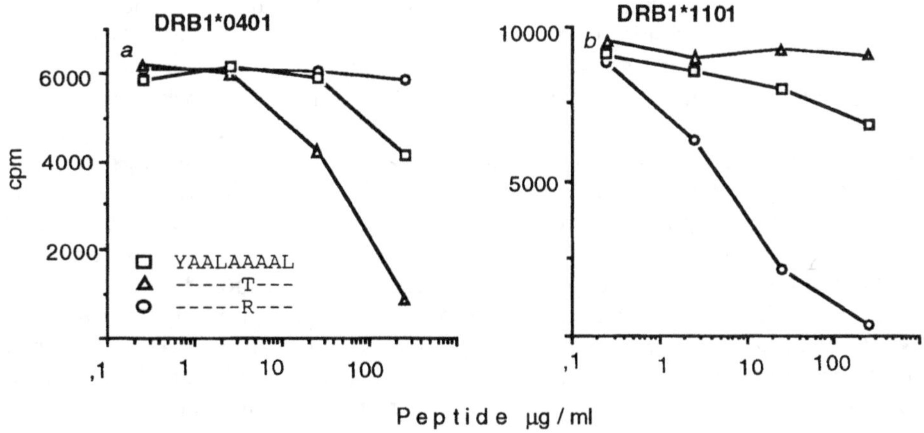

Figure 2. Effects on allele-specific binding of anchor substitutions in designer peptides.

Concluding Remarks

By screening a large number of peptide sequences expressed on the surface of filamentous phage for their ability to bind to RA-associated and nonassociated HLA-DR molecules we have identified binding motifs for three DR alleles. We identified promiscuous and allele specific anchors. In almost all DR binding peptides we found an aromatic amino acid followed, after two variable residues, by an amino acid with hydrophobic side chains. We defined the aromatic amino acid as position 1. A third anchor at position 6 was different for the three DR alleles: small residues for DRB1*0101, amino acids with hydroxyl groups for DRB*0401 and positively charged amino acids for DRB1*1101.

The recent structural determination of the peptide binding groove of murine H2D^b [5] and HLA-B27 [17] class I molecules has revealed the presence of conserved and polymorphic pockets of varying shapes that correspond to anchor residues of binding peptides. The identification of anchor residues in class II binding peptides suggest that specific pockets also exist in MHC class II molecules. Moreover, computer-aided modeling predicts that the alpha and the beta chain each form approximately half of the peptide binding cleft [7]. Unlike the beta chain, the alpha chain is constant in all DR molecules. Since the promiscuous anchor residues are situated in one half of the peptides (position 1 and 4), while the allele specific residues are in the second half of the peptides (position 6), it is tempting to speculate that the promiscuous anchor residues bind to the nonpolymorphic DRα chain, while the variable beta chain should bear polymorphic pockets for the variable anchor residues identified at position 6.

The knowledge of anchor position and other general rules for peptide-binding to MHC molecules should prove very useful for the design of MHC-specific antagonists. As discussed above, this strategy is directed at the development of a low molecular weight, possibly nonpeptidic antagonist that blocks the peptide binding site of disease-linked class II MHC molecules. Such an antagonist is expected to interfere with the presentation of antigenic peptide ligands including those involved in the activation of autoimmune T cells. The advantage of this strategy is that it is applicable without specific knowledge of the disease inducing autoantigen(s). Since different class II molecules exhibit different peptide binding specificities, the antagonist is expected to be selective, i.e., it should bind to the RA-linked DR molecules, but should not interfere with peptide binding to most other DR allotypes, or to additional class II isotypes, e.g., DP and DQ. Thus, a considerable part of the host's antigen presenting capacity should remain intact, and consequently, the patients would not be severely immunocompromised.

References

1. Todd JA, Acha-Orbea H, Bell JI, et al. A molecular basis for MHC class II-associated autoimmunity. Science (Washington, DC) 1988;240:1003-1009.
2. Wraith DC, Smilek DE, Mitchell DJ, Steinman L, McDevitt HO. Antigen recognition in autoimmune encephalomyelitis and the potential for peptide-mediated immunotherapy. Cell 1989;59:247-255.
3. Adorini L, Barnaba V, Bona C, et al. New perspectives on Immunointervention in autoimmune diseases. Immunol Today 1990;11:383-386.
4. Bjorkman PJ, Saper MA, Samraoui B, Bennet WS, Strominger JL, and Wily DC. Structure of the human class I histocompatibility antigen, HLA-A2. Nature 1987;329:506-512.
5. Fremont DH, Matsumura M, Stura EA, Peterson PA, and Wilson IA. Crystal structures of two viral peptides in complex with murine MHC class I H-2Kb. Science 1992;257:919-927.
6. Madden DR, Gorga JC, Strominger JL, and Wiley DC. The structure of

HLA-B27 reveals nonamer self-peptides bound in an extended conformation. Nature 1991;353:321-325.

7. Brown JH, Jardetzky T, Saper MA, Samraoui B, Bjorkman PJ, and Wiley DC. A hypothetical model of the foreign antigen binding site of class II histocompatibility molecules. Nature 1988;332:845-850.
8. Adorini L, Muller S, Cardinaux F, Lehmann PV, Falcioni F, and Nagy ZA. In vivo competition between self peptides and foreign antigens in T cell activation. Nature 1988;334:623-625.
9. Sakai K, Zamvil SS, Mitchell DJ, Hodgkinson S, Rothbard JB, and Steinman L. Prevention of experimental encephalomyelitis with peptides that block interaction of T cells with major histocompatibility complex protein. Proc Natl Acad Sci USA 1989;86:9470-9474.
10. Hurtenbach U, Lier E, Adorini L, and Nagy ZA. Prevention of autoimmune diabetes in non-obese diabetic mice by treatment with a class-II major histocompatibility complex-blocking peptide. J Exp Med 1993;177(5):1499-1504.
11. Hammer J, Takacs B, and Sinigaglia F. Identification of a motif for HLA-DR1 binding peptides using M13 display libraries. J Exp Med 1992;176:1007-1013.
12. Cwirla SE, Peters EA, Barret RW, and Dower WJ. Peptides on phage: A vast library of peptides for identifying ligands. Biochemistry 1990;87:6378-6382.
13. Devlin JJ, Paniganiban LC, and Devlin PE. Random peptide libraries: a source of specific protein binding molecules. Science 1990;249:404-406.
14. Scott JK and Smith GP. Searching for peptide ligands with an epitope library. Science 1990;248:386-390.
15. Parmeley SF and Smith GP. Antibody-selectable filamentous fd phage vectors: affinity purification of target genes. Gene 1988;73:305-309.
16. Hammer J, Valsasnini P, Tolba K, et al. Promiscuous and allele-specific anchors in HLA-DR binding peptides. Cell 1993, in press.
17. Madden DR, Gorga JC, Strominger JL, and Wiley DC. The three-dimensional structure of HLA-B27 at 2.1 A resolution suggests a general mechanism for tight peptide binding to MHC. Cell 1992;70:1035-1048.

THE THERAPEUTIC USE OF ANTI-CD3 MONOCLONAL ANTIBODIES IN AUTOIMMUNE DISEASES-APPLICATION TO TYPE I INSULIN-DEPENDENT DIABETES

Lucienne Chatenoud, Eric Thervet and Jean-François Bach
Inserm U25
Hôpital Necker
161 Rue de Sèvres
75015 Paris
FRANCE

Introduction

Type I insulin dependent diabetes mellitus (IDDM) is an autoimmune disease [1]. The non-obese diabetic (NOD) mouse, originally described by Makino et al. is an accurate experimental model of human IDDM; overt metabolic disturbances (glycosuria and hyperglycemia), spontaneously appear by 15 weeks of age and are preceded by a progressive infiltration of the islets of Langerhans with mononuclear cells termed "insulitis" [2]. Insulitis appears by 5 weeks of age and is observed both in NOD males and females. Depending on the colony, the cumulative incidence of overt IDDM reaches 60-80% in females and 10-80% in males. There is compelling evidence confirming the essential role of T cell mediated immunity in the pathogenesis of IDDM: 1) neonatal thymectomy prevents the onset of disease [3] and athymic NOD nude mice are free of insulitis and IDDM [4], 2) IDDM can be adoptively transferred to NOD neonates, irradiated adult NOD mice and more recently to athymic NOD nude mice, by injection of splenic T cells from diabetic animals; both CD4+ and CD8+ lymphocytes are needed to achieve effective transfer of disease [4-6], 3) different immunointervention strategies specific for T cells are effective in preventing IDDM.

A number of immunosuppressive agents have been shown to prevent IDDM in NOD mice when administered before the onset of the insulitis and/or the clinical manifestations. These include cyclosporin A [7] and different monoclonal antibodies (MoAb) such as, anti-CD3 [8], anti-CD4 [9], anti-CD8 [10], anti-T cell receptor (TcR) $\alpha\beta$ [11], anti-CD11d [12] and MoAb to class II antigens of the major histocompatibility complex [13]. In contrast, the treatment of mice with established metabolic disease, which is the situation encountered in clinical practice, still represents a major challenge. In addition, most therapies listed above, both in

A. G. Dalgleish et al. (eds.), The Impact of Biotechnology on Autoimmunity, 123–127.

mice and humans required continuous administration to maintain their effect, thus constituting a hardly acceptable risk for young patients with recent onset IDDM. This is further illustrated by the fact that therapy with anti-TcRαβ at high doses could reverse established spontaneous IDDM but the effect was transient and disease recurred when the MoAb was cleared [11]. It thus appeared of the utmost importance to look for short-term therapeutic procedures that could lead to long-term remission of established disease. A recent report by Maki et al. [14] showed that stable remission of overt IDDM could be achieved by complete *in vivo* T cell debulking, as induced by treatment with a rabbit anti-lymphocyte polyclonal antiserum or a mixture of depleting anti-CD4 and anti-CD8 MoAbs at high doses (~ 5mg cumulated MoAb dose/mouse), applied within 14 days of disease onset [14]. However, in Maki's study, the profound and prolonged T cell depletion that was required to achieve the long-term remission totally precludes the applicability of this protocol to the clinical situation. In addition, in this study diabetic mice also received insulin which may have contributed to the preservation of remnant β cells; it is well established that insulin therapy can promote remission in human IDDM [15].

The Immunosuppressive Capacity of Anti-CD3 Antibodies in NOD Mice

Anti-CD3 MoAb are potent immunosuppressants that essentially act by reversibly clearing CD3/TCR complexes due to the antigenic modulation they promote [16-18]. They are particularly effective in both preventing and/or reversing acute rejection of organ allografts, a capacity extensively documented both in the clinical and the experimental setting [16,19-20]. Low-dose anti-CD3 treatment is thus particularly adapted to short-term immunosuppression as it avoids the long-term T cell depletion observed with immunosuppressive MoAbs, including some anti-CD4 species, such as GK 1.5 [9]. We first determined the dose of anti-CD3 to apply to adult male and female NOD mice of various ages; a dose of 5 μg was well tolerated, despite the systemic release of tumor necrosis factor (TNF). We have shown the poor tolerance, in adult mice of most strains, of the first anti-CD3 injections is related to a massive cell activation and the cytokine release it promotes [21]. The immunosuppressive efficacy of this low-dose anti-CD3 treatment was studied by testing its capacity to influence spontaneous insulitis in NOD mice. In our breeding colony spontaneous insulitis develops by 5 to 8 weeks and by 10 weeks of age more than 80% of males and females show signs of insulitis. However, at 25 weeks only 10-15% of males had become diabetic, compared with 50-60% of females.

A 5-day course of anti-CD3 (5μ/day) in 10-week-old NOD females induced a major reduction of the insulitis but the effect was transient, disappearing within two weeks of the last injection.

This short anti-CD3 course also prevented the development of cyclophosphamide (CY)-induced IDDM. CY, an alkylating agent known to augment certain immune responses such as delayed-type hypersensitivity reactions, accelerates and/or promotes both in females and males the onset of IDDM [22-24]. On the basis of the data reported by several groups, this effect is interpreted as secondary to the elimination of down-regulatory cells. It is important to stress that CY-induced diabetes may be prevented by adoptive transfer of mononuclear cells from nondiabetic NOD mice [24]. The 5-day anti-CD3 treatment started one day prior to the second CY injection--when approximately 15 to 20% of mice were already diabetic--was highly effective both in preventing progression to IDDM and in reversing overt disease.

Anti-CD3 Treatment of Mice Presenting Established Diabetes

The same short-term low-dose treatment protocol was used to determine whether anti-CD3 could stop the progression of spontaneous ongoing autoimmune IDDM. Overt diabetes in NOD females was identified by twice-weekly screening for glycosuria; when glycosuric, mice were tested for hyperglycemia. Mice showing stable hyperglycemia (10 to 34 weeks of age) were randomized to receive short-term low-dose treatment with either anti-CD3 (5 μg/day for 5 consecutive days) or normal hamster immunoglobulins. Progressive remission of established disease (defined as the disappearance of glycosuria and a return to normal glycemia) occurred in 64-80% of anti-CD3-treated mice within 2 to 4 weeks of the last MoAb injection (L. Chatenoud et al., submitted). The event was significantly more frequent than spontaneous remission in control NOD mice treated with the irrelevant hamster immunoglobulins. Insulitis was evidenced in mice showing anti-CD3-induced remission but it did not present the typical invasive-destructive pattern typical of NOD females with progressive diabetes. Immunoenzymatic staining of frozen pancreatic tissue sections showed that the insulitis observed in mice showing anti-CD3-induced remission essentially included CD3+ TCRαβ CD4+ or CD8+ T lymphocytes. Mice showing anti-CD3-induced remission were not globally immunocompromised: they rejected normally skin allografts but did not destroy syngeneic islet grafts, further stressing the specificity of the anti-CD3-induced remission for β cell-associated antigens.

These results provide the first indication that a short course of anti-CD3 given alone can restore self-tolerance in adult mice with established autoimmune diabetes. The mechanisms leading to the establishment of such long-term remission are presently being analyzed. Extrapolating from other experimental situations [25-30], it is tempting to speculate that T cell anergy or active T cell-mediated suppressor circuits are active in our model. It is also possible that anti-CD3-induced T cell activation participates in this effect.

References

1. Castano L, Eisenbarth GS. Type I diabetes: a chronic autoimmune disease of human, mouse and rat. Ann Rev Immunol 1990;8:647-679.

2. Makino S, Kunimoto K, Muraoka Y, Mizushima Y, Katagiri K, Tochino Y. Breeding of a non-obese diabetic strain. Exp Anim 19080;29:1-13.
3. Ogawa M, Maruyama T, Hasegawa T. The inhibitory effect of neonatal thymectomy on the incidence of insulitis in NOD mice. Biomed Research 1985;6:103-108.
4. Yagi H, Matsumoto M, Kunimoto K, Kawagushi J, Makino S, Harada M. Analysis of the roles of CD4+ and CD8+ T cells in autoimmune diabetes of NOD mice using transfer to NOD athymic nude mice. Eur J Immunol 1992;22:2387-2393.
5. Bendelac A, Carnaud C, Boitard C, Bach JF. Syngeneic transfer of autoimmune diabetes from diabetic NOD mice to healthy neonates. Requirement of both L3T4+ and Lyt-2+ T cells. J Exp Med 1987;166:823-832.
6. Wicker LS, Miller BJ, Mullen Y. Transfer of autoimmune diabetes mellitus with splenocytes from non-obese diabetic (NOD) mice. Diabetes 1986;35(8):855-86.
7. Mori Y, Suko M, Okudiara H, Matsuba I, Tsuruoka S, Sasaki A, Yokoyama H, Tanase T, Shida T, Nishimura M, Terada E, Ikeda Y. Preventive effect of cyclosporin on diabetes in NOD mice. Diabetologia 1986;29:244-247.
8. Hayward AR, Shreiber M. Neonatal injection of CD3 antibody into nonobese diabetic mice reduces the incidence of insulitis and diabetes. J Immunol 1989;143:1555-1559.
9. Shizuru JA, Taylor-Edwards C, Banks BA, Gregory AK, Fathman CG. Immunotherapy of the nonobese diabetic mouse: treatment with an antibody to helper T cells. Science 1988;240:659-662.
10. Charton B, Bacelj A, Mandel TE. Administration of silica particles or anti-Lyt-2 antibody prevents b cell destruction in NOD mice given cyclophosphamide. Diabetes 1988;37:930-935.
11. Sempé P, Bédossa P, Richard MF, Villà MC, Bach JF, Boitard C. Anti α/β T cell receptor monoclonal antibody provides an efficient therapy for autoimmune diabetes in nonobese diabetic (NOD) mice. Eur J Immunol 1991;21:1163-1169.
12. Hutchings P, Rosen H, O'Reilly L, Simpson E, Gordon S, Cooke A. Transfer of diabetes in mice prevented by blockade of adhesion-promoting receptor on macrophages. Nature 1990;348:639-642.
13. Boitard C, Bendelac A, Richard MF, Carnaud C, Bach JF. Prevention of diabetes in non-obese diabetic mice by splenic T cells. Proc Natl Acad Sci (USA) 1988;85:9719-9725.
14. Maki T, Ichikwa T, Blanco R., Porter J. Long-term abrogation of autoimmune diabetes in nonobese diabetic mice by immunotherapy with anti-lymphocyte serum. Proc Natl Acad Sci USA 1992;89:3434-3438.
15. Shah CS, Malone JI, Simpson NE. A randomized trial of intensive insulin therapy in newly diagnosed insulin-dependent diabetes mellitus. N Engl J

Med 1989;320:550-554.

16. Hirsch R, Eckhaus M, Auchincloss Jr H, Sachs DH, Bluestone JA. Effect of in vivo administration of anti-T3 monoclonal antibody on T cell function in mice. I. Immunosuppression of transplantation responses. J Immunol 1988;140:3766-3772.
17. Chatenoud L, Bach JF. Antigenic modulation - A major mechanism of antibody action. Immunol Today 1984;5:20-25.
18. Chatenoud L, Baudrihaye MF, Kreis H, Goldstein G, Schindler J, Bach JF. Human in vivo antigenic modulation induced by the anti-T cell OKT3 monoclonal antibody. Eur J Immunol 1982;12:979-982.
19. Ortho Multicenter Transplant Study Group. A randomized clinical trial of OKT3 monoclonal antibody for acute rejection of cadaveric renal transplants. New Engl J Med 185;313:337-342.
20. Debure A, Chkoff N, Chatenoud L, Lacombe M, Campos H, Noel LH, Goldstein G, Bach JF, Kreis H. One month prophylactic use of OKT3 in cadaver kidney transplant recipients. Transplantation 1988;45:546-553.
21. Ferran C, Dy M, Sheehan K, Merite S, Schreiber R, Landais P, Grau G, Bluestone J, Bach JF, Chatenoud L. Inter-mouse strain differences in the in vivo anti-CD3 induced cytokine release. Clin Exp Immunol 1991;86:537-543.
22. Harada M, Makino S. Promotion of spontaneous diabetes in non-obese diabetes-prone mice by cyclophosphamide. Diabetologia 1984;27:604-606.
23. Yasunami R, Bach JF. Anti-suppressor effect of cyclophosphamide on the development of spontaneous diabetes in NOD mice. Eur J Immunol 1988;18:481-484.
24. Charlton B, Bacelj A, Slattery RM, Mandel TE. Cyclophosphamide-induced diabetes in NOD/WEHI mice: Evidence for suppression in spontaneous autoimmune diabetes mellitus. Diabetes 1989;38:441-447.
25. Cobbold SP, Qin S, Waldmann H. Reprogramming the immune system for tolerance with monoclonal antibodies. Seminars Immunol 1990;2:377-386.
26. Qin S, Cobbold S, Benjamin R, Waldmann H. Induction of classical transplantation tolerance in the adult. J Exp Med 1989;169:779-794.
27. Alters SE, Shizuru JA, Ackerman J, Grossman D, Seydel Y, Fathman G. Anti-CD4 mediates clonal anergy during transplantation tolerance induction. J Exp Med 191;173:491-494.
28. Alters SE, Grossman D, Fathman G. Anti-CD4-mediated transplantation tolerance: mechanism and memory. In: Burlingham WJ, eds. A critical analysis of monoclonal antibody therapy. CRC Press, Boca Raton: p.79-99.
29. Qin S, Cobbold SP, Heather P, Elliott J, Kioussis D, Davies J, Waldmann H. "Infectious" transplantation tolerance. Science 1993;259:974-977.
30. Dallman M, Shiho O, Page TH, Wood KJ, Morris PJ. Peripheral tolerance to alloantigen results from altered regulation of the interleukin 2 pathway. J Exp Med 1991;173:79-87.

ENGINEERING OF IL-1 ANTAGONISTS

Diana Boraschi and Aldo Tagliabue
Dompé S.p.A
L'Aquila
ITALY

Introduction

Interleukin 1 (IL-1) is a family of pleiotropic cytokines, produced mainly by stimulated mononuclear phagocytes, which play a pivotal role in the onset and development of host defense reactions [1,2]. The involvement of IL-1 in the initiation and amplification of defense responses, both immunological and inflammatory, is in fact of major importance for the host reaction mechanisms to invasion and infection. However, because of its direct and indirect inflammatory and immunostimulatory activities, IL-1 is also responsible for many of the destructive effects of chronic inflammatory diseases and arthralgias (including rheumatoid arthritis); it is involved in the altered immunological responses of allergy and autoimmunity; and it eventually takes a key part in the lethal septic shock syndrome [3,4].

Given the importance of IL-1 in these pathological events, antagonism of IL-1 action has become the object of intense investigation, as a promising therapeutic avenue of focused and effective intervention. Indeed, a natural antagonist of IL-1 has been recently identified, i.e. the third member of the IL-1 family (besides IL-1α and IL-1β) which shares with IL-1 the overall structure and conformation and the ability to bind to IL-1 receptors (IL-1R) but is devoid of biological activities [5]. This receptor antagonist (IL-1ra) thus occupies with high affinity the IL-1R making them unavailable for binding active IL-1α and IL-1β. Another naturally occurring IL-1 antagonist has been also identified [6,7], that is the extracellular part of the IL-R_{II}, the second type of IL-1R which apparently is unable to mediate activation signals [8]. The soluble IL-R_{II} (sIL-1R_{II}) is released from the cell membrane without losing its high binding capacity for IL-1β and acts by capturing IL-1 in the microenvironment before it can bind to membrane active receptors (IL-1R_I) and trigger them.

In the present paper, the strategies for development of effective IL-1 antagonists will be examined. These start from the information obtained from structure and activity of the naturally occurring antagonists, implemented with information on the

A. G. Dalgleish et al. (eds.), The Impact of Biotechnology on Autoimmunity, 129–134.

structure-function relationship within the active IL-1 protein, in the attempt to construct a new molecule with enhanced binding characteristics and possessing or not chosen biological activities

Results and Discussion

Four types of possible IL-1 antagonists could be envisaged, which can be active in inhibiting IL-1 functions both *in vitro* and *in vivo:*

1. IL-1ra and its mutants
2. IL-1R_I and IL-R_{II} soluble forms
3. Antibodies to IL-1R
4. Mutants of IL-1

The first two are based on the prototype IL-1 antagonists, i.e. the naturally occurring IL-1ra and sIL-1R, with introduction of modifications to improve their binding capacity and their antagonistic effect. As a third possibility, antibodies which block and down-regulate IL-1R can be an effective way of inhibiting IL-1 effects. Finally, mutations of IL-1 itself can be designed, such as to lose selectively some or all IL-1-like biological effects though maintaining the capacity to occupy IL-1R. These four classes of possible inhibitors will hereafter be examined in more detail.

IL-1ra AND ITS MUTANTS

The structure of IL-1ra is very similar to that of active IL-1 molecules (IL-1α and IL-1β), with which IL-1ra shares the ability to bind with high affinity to IL-R_I. On the other hand, IL-1ra has very low affinity for the IL-1R_{II}, at variance with IL-1β. Thus, IL-1ra is very effective in binding to the activating receptor IL-1R_I but incapable of occupying the inhibitory receptor IL-1R_{II}. Contrary to IL-1α and IL-1β, IL-1ra cannot stimulate cell activation upon receptor binding and therefore acts as IL-1 antagonist at the receptor level [9,10]. The excellent capacity of IL-1ra of inhibiting IL-1 effects *in vitro* has been confirmed by very exciting results *in vivo*, where IL-1ra can save animals from death following endotoxin shock [11]. However, its use in humans in septic shock, after encouraging preliminary results, does not appear to be as promising as hoped, with the need of exceedingly high doses to see a scarcely significant increase in survival rates. This result is possibly based on the mechanism of cell activation by IL-1, where very few receptors apparently need to be occupied to bring about full activation of cellular functions [12]. Therefore, an effective antagonist at the receptor level should occupy all the IL-1R on the cell surface (since few free receptors are sufficient for full IL-1 action) with a very high affinity to avoid displacement by active IL-1. Thus, modifications of the structure of IL-1ra can be designed, in order to increase its affinity for IL-1R_I (augmenting the association constant and decreasing the dissociation constant), based on optimization of the predicted sites of receptor interaction and comparison with the structure of IL-1. It is important, in designing optimized IL-1ra, to maintain the inability of the molecule to bind IL-1R_{II}, since this receptor in its soluble form is very effective in capturing and inhibiting IL-1 and, if occupied by IL-1ra, would lose its inhibiting potential and at the same time inhibit the IL-1ra effect.

IL-1R_I AND IL-1R_{II} SOLUBLE FORMS

IL-1R_I and IL-R_{II}, the two types of IL-1R presently known, are single chain membrane receptors belonging to the immunoglobulin superfamily of proteins. The extracellular parts of IL-1R_I and of IL-1R_{II} are composed by three Ig-like domains closed by disulfide bonds and share a significant degree of structural and sequence homology [8]. On the other hand, the two receptors are very different in their intracellular domain, since that of IL-1R_{II} is extremely short and unlikely to take part in any mechanism of signal transduction. Thus, since the mechanisms of cell activation through this type of IL-1R have remained elusive, it is generally believed that the main role of IL-1R_{II} is that of acting as a "decoy" receptor able to capture IL-1 and therefore to avoid its interaction with the activating IL-1R_I [13]. Indeed, a naturally occurring IL-1 inhibitor has been identified, which is the soluble form of IL-1R_{II}, abundantly produced by many cell types (mainly B cells, macrophages and PMN). The IL-1R_{II} extracellular domain is released from the cell surface and can act as a soluble inhibitor of IL-1. The affinity of the B cell-derived soluble receptor is very good for IL-1β (which represent the majority of IL-1 in solution), whereas it is almost unable to bind IL-1α (the IL-1 type which is mostly cell-associated) and IL-1ra [6]. Thus, the sIL-1R_{II} could be an excellent inhibitor of IL-1 activities which does not interfere with the other natural inhibitory molecule IL-1ra. Conversely, the extracellular domain of IL-1R_I is not released naturally and it does not exist as a natural soluble inhibitor. However, it has been constructed by genetic engineering and used to inhibit IL-1 effects with good results in experimental models [14]. The problem of sIL-1R_I could be that this receptor binds well not only to active IL-1 types (IL-1α and IL-1β) but also to the inhibitory IL-1ra. Thus the overall efficacy of sIL-1R_I can be diminished by its ability to neutralize the natural antagonist and to be less available to neutralize active IL-1.

ANTIBODIES TO IL-1R

The use of antibodies, specifically able to neutralize and downregulate one type of

IL-1R, can be envisaged as a possible way of effectively inhibit IL-1 activities. The IL-1 inhibiting capacity of anti-murine IL-1R_I antibodies has been extensively tested *in vivo* in the mouse with interesting results, although their efficacy in inhibiting IL-1-dependent pathologies (e.g. endotoxin shock) has not been assessed [15]. A monoclonal antibody generated to the human IL-1R_{II} has been used in the mouse and in the rat with results which indicate the possibility that in fact also the second type of IL-1R could have an activation role, beside its downregulation role, at least in some anatomical districts [16]. The ALVA-42 antibody inhibits binding of IL-1 to IL-1R_{II} but not to IL-1R_I and it can block a series of IL-1 activities, such as bone resorption, pyrogenicity and thermogenesis, anorexia, immunostimulation, neutrophilia, decrease in gastric acidity, whereas it is without effect on other IL-1 effects (e.g. hypoglycemia, increase in plasma cortico-sterone and IL-6). Some of the IL-1 effects which are inhibited by the anti-IL-1R_{II} antibody are also decreased by anti-IL-1R_I antibodies, indicating that to the outcome of the IL-1 effect *in vivo* participate both cells bearing IL-1R_I and cells with IL-1R_{II}, or cells bearing both receptors together. In other cases (e.g. anorexia), the inhibitory effect could only be seen with the anti-IL_{II} antibody, indicating the main involvement of this receptor in this central effect of IL-1. To stress the importance of the IL-1R_{II} in mediating some IL-1 activities, administration of ALVA-42 antibodies could significantly inhibit endotoxin shock and death in the mouse, similar to what is observed with IL-1ra. This indicates therefore that both IL-1R_I (inhibited by IL-1ra) and IL-1r_{II} (inhibited by ALVA) mediated the toxic effects of IL-1 during shock.

MUTANTS OF IL-1

As the last approach for designing IL-1 antagonists, mutants of IL-1 can be designed which maintain the ability of binding IL-1R but selectively lose some or all the IL-1 activities. These mutants would therefore mimic IL-1ra (if devoid of all IL-1 activities) or represent a sort of hybrid IL-1/IL-1ra capable of selectively inhibiting some IL-1 effects and of triggering other IL-1 actions. However, IL-1 mutants which have been so far constructed with this scope usually show a strict correlation between capacity to bind IL-1R_I and ability to exert biological activities. In some cases, preliminary results would show constructs which maintain the binding capacity while losing the biological activity (e.g. in IL-1β substitution of C in position 187 with S or of D in position 261 with K) [17,18], or mutants with unaltered binding capacity and maintenance of only some biological activities (e.g. in IL-1β substitution of R in position 127 with G) [19]. Interestingly, it is apparently possible to dissociate between binding capacity for IL-1R_I and IL-1R_{II} and thus to obtain mutants of IL-1 capable of binding only to one of the two IL-1R (e.g. IL-1β mutants with substitutions L to A in 122, or R to K or to A in 127, or K to S in 219, or E to S in 221 fully bind to IL-1R_{II} but lose the capacity of binding to IL-1R_I) [20]. These latter mutants should conceivably be devoid of IL-1 biological activities dependent on triggering of IL-1R_I (since unable to bind to this receptor) but, being capable of occupying the "decoy" receptor IL-1R_{II}, they should function as inhibitors of the natural inhibitor sIL-1R_{II}. In addition, if IL-1R_{II} is indeed involved in some of the IL-1 biological effects, these mutants would be able to agonize IL-1 for such effects. Assay of their agonist and antagonist activity *in vitro* and *in vivo* will clarify their selectivity and indicate their possible therapeutic application.

References

1. Dinarello CA. Interleukin-1 and its biologically related cytokines. Adv Immunol 1989;44:153-205.
2. Dinarello CA. Interleukin-1 and interleukin-1 antagonism. Blood 1991;77:1627-1652.
3. Okusawa S, Gelfand A, Ikejima T, Connolly J, Dinarello CA. Interleukin-1 induces a shock-like state in rabbits. Synergism with tumor necrosis factor and the effect of cyclooxygenase inhibition. J Clin Invest 1988;81:1162-1172.
4. Waage A, Brandtzaeg P, Halstensen A, Kierulf P, Espevik T. The complex pattern of cytokines in serum from patients with meningococcal septic shock. Association between interleukin 6, interleukin 1, and fatal outcome. J Exp Med 1989;169:333-338.
5. Arend WP. Interleukin 1 receptor antagonist. A new member of the interleukin-1 family. J Clin Invest 1991;88:1445-1451.
6. Symons JA, Eastgate JA, Duff GW. Purification and characterization of a novel soluble receptor for interleukin 1. J Exp Med 1991;174:1251-1254.
7. Giri JG, Newton RC, Horuk R. Identification of soluble interleukin-1 binding protein in cell-free supernatants. J Biol Chem 1990;265:17416-17419.
8. McMahan CJ, Slack JL, Mosley B, et al. A novel IL-1 receptor, cloned from B cells by mammalian expression, is expressed in many cell types. EMBO J 1991;10:2821-2832.
9. Dripps DJ, Brandhuber BJ, Thompson RC, Eisenberg SP. Interleukin-1 (IL-1) receptor antagonist binds to the 80-kDa IL-1 receptor but does not initiate IL-1 signal transduction. J Biol Chem 1991;266:10331-10336.
10. Dinarello CA, Thompson RC. Blocking IL-1: interleukin 1 receptor antagonist *in vivo* and *in vitro.* Immunol Today 1991;11:404-410.
11. Ohlsson K, Bjork P, Bergenfeldt M, et al. Interleukin-1 receptor antagonist reduces mortality from endotoxin shock. Nature 1990;348:550-552.
12. Dinarello CA, Clark BD, Puren AJ, Savage N, Rosoff PM. The interleukin 1 receptor. Immunol Today 1989;10:49-51.
13. Colotta F, Re F, Muzio M, et al. The type II IL-1 "receptor" as a decoy target for IL-1 in polymorphonuclear leukocytes. Science 1993; in press.
14. Fanslow WC, Sims JE, Sassenfeld H, et al. Regulation of alloreactivity in vivo by a soluble form of the interleukin-1 receptor. Science 1990;248:739-

742.

15. McIntyre KW, Stepan GJ, Kolinsky KD, et al. Inhibition of interleukin 1 (IL-1) binding and bioactivity in vitro and modulation of acute inflammation in vivo by IL-1 receptor antagonist and anti-IL receptor monoclonal antibody. J Exp Med 1991;173:931-939.
16. Boraschi D, Ghiara P, Armellini D, Tagliabue A. What is the second IL-1 receptor for? In: Bartfai T, Ottoson D, editors. Neuro-immunology of fever. Oxford: Pergamon Press, 1992:19-28.
17. Folliard F, Touchet N, Terlain B. Comparison of the binding of IL-1 and mutants on lymphocytes and chondrocytes and their effects on some biological activities. Discovery of an IL-1β mutant with dissociated properties. Cytokine 1990;S1:51.
18. Ju G, Labriola-Tompkins E, Campen CA, et al. Conversion of the interleukin 1 receptor antagonist into an agonist by site-specific mutagenesis. Proc Natl Acad Sci USA 1991;88:2658-2662.
19. Gehrke L, Jobling SA, Paik LSK, McDonald B, Rosenwasser LJ, Auron PE. A point mutation uncouples human interleukin-1β biological activity and receptor binding. J Biol Chem 1990;265:5922-5925.
20. Labriola-Tompkins E, Chandran C, Kaffka KL, et al. Identification of the discontinuous binding site in human interleukin 1β for the type I interleukin 1 receptor. Proc Natl Acad Sci USA 1991;88:11182-11186.

Index

Medical Science Symposia Series

1. A. M. Gotto, C. Lenfant, R. Paoletti (eds.) and M. Soma (ass.ed.): *Multiple Risk Factors in Cardiovascular Disease.* 1992 ISBN 0-7923-1938-9
2. A. L. Catapano, A. M. Gotto, Jr., L. C. Smith and R. Paoletti (eds.): *Drugs Affecting Lipid Metabolism.* 1993 ISBN 0-7923-2232-0
3. T. Godfraind, S. Govoni, R. Paoletti and P. M. Vanhoutte (eds.): *Calcium Antagonists. Pharmacology and Clinical Research.* 1993 ISBN 0-7923-2259-2
4. D. Galmarini, L. R. Fassati, R. Paoletti and S. Sherlock (eds.): *Drugs and the Liver: High Risk Patients and Transplantation.* 1993 ISBN 0-7923-2307-6
5. P. M. Vanhoutte, P. R. Saxena, R. Paoletti, N. Brunello (eds.) and A. S. Jackson (ass.ed.): *Serotonin. From Cell Biology to Pharmacology and Therapeutics.* 1993 ISBN 0-7923-2518-4
6. A. G. Dalgleish, A. Albertini and R. Paoletti (eds.): *The Impact of Biotechnology on Autoimmunity.* 1994 ISBN 0-7923-2724-1

KLUWER ACADEMIC PUBLISHERS – DORDRECHT / BOSTON / LONDON

GPSR Compliance
The European Union's (EU) General Product Safety Regulation (GPSR) is a set of rules that requires consumer products to be safe and our obligations to ensure this.

If you have any concerns about our products, you can contact us on

ProductSafety@springernature.com

In case Publisher is established outside the EU, the EU authorized representative is:

Springer Nature Customer Service Center GmbH
Europaplatz 3
69115 Heidelberg, Germany

www.ingramcontent.com/pod-product-compliance
Lightning Source LLC
LaVergne TN
LVHW011805200926
848484LV00003B/784
9780792327240